Transient Control
of Gasoline
Engines

Transient Control of Gasoline Engines

Tielong Shen • Jiangyan Zhang

Xiaohong Jiao • Mingxin Kang

Junichi Kako • Akira Ohata

CRC Press
Taylor & Francis Group
Boca Raton   London   New York

CRC Press is an imprint of the
Taylor & Francis Group, an **informa** business

CRC Press
Taylor & Francis Group
6000 Broken Sound Parkway NW, Suite 300
Boca Raton, FL 33487-2742

© 2016 by Taylor & Francis Group, LLC
CRC Press is an imprint of Taylor & Francis Group, an Informa business

No claim to original U.S. Government works

Printed on acid-free paper
Version Date: 20150914

International Standard Book Number-13: 978-1-4665-8426-6 (Hardback)

Visit the Taylor & Francis Web site at
http://www.taylorandfrancis.com

and the CRC Press Web site at
http://www.crcpress.com

We dedicate this book to our families and parents.

Contents

List of Figures

List of Tables

Preface

In the past three decades, internal combustion engines have become computerized machines as many actuators developed during the decades have to be controlled by the electrical control unit (ECU). Roughly speaking, a modern combustion engine cannot be operated without the ECU control, and the main purpose of the control is to improve the engine performance parameters, such as fuel efficiency, combustion quality, emissions, and so on. Recently, engine control focuses not only on the static operating mode but also on the transient operating conditions, including acceleration, load changes, cold or warm starting, coordinating operations with alternative power sources.

Indeed, in the past, the strategy to manage combustion engines was usually focused on the static- or pseudo-static mode even under ECU control, because it is very difficult to handle combustion engines at the transient mode due to the strong nonlinearity and uncertainties in the combustion event, thermodynamics, fluid dynamics in the inlet and exhaust paths and fuel paths, and so on. In general, in thermal engineering, the attention is focused on stable combustion and efficiency improvement evaluated in the sense of cyclic average. Dealing with the combustion engine as a dynamical system and approaching the goal of efficiency improvement with advanced control theory have recently been spotlighted. Benefitting from the rapid progress in car electronics and digital processing technology and motivated by this trend, modeling and control of combustion engines has become a very active field in control engineering. The intent of this book is to present a model-based transient control design methodology from the perspective of dynamical system control theory. For this purpose, several control issues are addressed concerning transient operation, cycle-to-cycle transient, and cylinder-to-cylinder balancing. As will be shown, it is essential to explore a control-oriented model according to the control objective for exploiting advanced control theory in transient control strategies and to combine the analysis on the engine physics with statistical methodology for constructing a control-oriented mode. A feature of the controller design process that has been explained in this book has three phases: modeling, the mode-based control law derivation, and test bench-based validation.

This book is organized as follows: An introduction to modeling, control design, and test bench is given in Chapter 1, highlighting an example of transient engine behavior and the control issues that have been addressed in the following chapters. Chapter 2 describes the dynamics issues, including thermal dynamics, fluid dynamics of the intake and fueling path, the rotational dynamics, and the heat release. How to simplify the model according to the control objective is depicted by showing a mean-value treatment of torque generation and air charge of individual cylinders. Chapter 3 introduces a design example of the problem of speed control. The control issues of idling operating mode and the starting transient mode are discussed. Chapter 4 focuses on the air–fuel ratio control problem. Model-based predictive feedforward compensation issues are investigated at first. Then, Lyapunov-based adaptive control scheme is presented to cope with the parametric uncertainties in the engine models. Chapter 5 is devoted to the real-time optimizing control problem, so-called receding horizon optimization, for torque tracking control and speed control. With regard to industrial application, this control principle is also called model predictive control as the optimal control strategy is decided along the model-based predicted trajectories. Parameter tuning and uncertainties are also discussed in the chapter. Imbalance in an individual cylinder is an important issue when transient control is targeted in a multi-cylinder engine. Chapter 6 focuses on cylinder-to-cylinder balancing control issues. The air–fuel ratio is chosen as evaluating index of the imbalance, and two control strategies, the model-based predictive control and the periodic time-varying model-based learning control, are described. As is well-known, an in-cylinder combustion event exhibits a stochastic property which leads to the statistical treatment of engine management. Chapter 7 provides a basic concept that tries to more actively deal with the stochastic property. By focusing on the cyclic residual gas fraction, the Markovian property of the residual gas fraction is investigated based on the experiment on the targeted engine, and then the problem of optimal control is solved in the sense of stochastic optimization with respect to the Markov chain model. Chapter 8 involves three benchmark problems regarding modeling and control of gasoline engines which have been raised by the last two authors of this book. The three benchmark problems, engine start control, identification of the engines, and the boundary modeling and extreme condition control, have been proposed by the researchers from the sense of industrial powertrain development, and it is believed that these practical, challenging issues would attract further challenges from the readers this book targets, especially the postgraduate students.

It should be noted that this book is the outcome of the authors' research activities of many years. An important feature of the research activities is the collaboration between the academia and industry. The first four authors are with the Department of Engineering and Applied Sciences, Sophia University, Japan, and the last two members of the second group of authors are experts from the Future Project Division of Toyota Motor Corporation, Japan. This collaboration is a long-term joint project between the two groups. As the leading author, I would like to list the author names here with gratitude, as some of them have left Sophia University: Dr. Jianyan Zhang, assistant professor of Dalian Nationalities University, China, who left Sophia in 2012 after having worked for this project for more than seven years; Dr. Xiaohong Jiao, professor of Yanshan University, China, who left Sophia University in 2004 and collaborated till now; Dr. Mingxin Kang who is currently serving as a research associate in Sophia University; and Mr. Akira Ohata and Mr. Junichi Kako from Toyota Motor Corporation, Japan, who are still continuing their creative collaboration with the academic community.

We are grateful to the effort and the prior work of our former students who received their PhD degrees under the supervision of the first author, in particular, Dr. Po Li, Dr. Munan Hong, Dr. Yinhua Liu, Dr. Jun Yang, Dr. Kenji Suzuki, and Dr. Tomoyuki Hara. We also thank Mr. Kota Sata of Toyota Motor Corporation for his friendly collaboration over the years and Mr. Yasufumi Oguri for his continuous support to the experimental research activities and maintenance of the test bench.

Finally, we thank our families for their continuous and unflinching support.

Prof. Dr. Tielong Shen
Sophia University, Tokyo, Japan

Akira Ohata
Toyota Motor Corporation, Shizuoka, Japan

Authors

Tielong Shen earned his PhD in mechanical engineering from Sophia University, Tokyo, Japan. From April 1992, he has been a faculty member and the chair of control engineering in the Department of Mechanical Engineering, Sophia University, where he currently serves as a full professor. Since 2005, he has also concurrently served as "Luojia Xuezhe" Chair Professor of Wuhan University, Wuhan, China. His research interests include control theory and applications in mechanical systems, powertrain, and automotive systems. He has authored and coauthored more than 160 journal articles and 9 books. He is presently also serving as a member of the IEEE Technical Committee on Automotive Control and IFAC Technical Committee on Automotive Control.

Jiangyan Zhang received BE and ME degrees in electrical engineering from Yanshan University, Qinhuangdao, China, in 2005 and 2008, respectively, and PhD in mechanical engineering from Sophia University, Tokyo, Japan, in 2011. From April 2011 to March 2013, she was a postdoctoral research fellow at SHEN Laboratory of Sophia University, and currently is an assistant professor with the College of Electromechanical and Information Engineering of Dalian Nationalities University, Dalian, China. Her research interests includes nonlinear dynamical control theory and applications to the automotive powertrain systems.

Xiaohong Jiao earned her PhD in mechanical engineering from Sophia University, Tokyo, Japan, in 2004. She is a professor with the Institute of Electrical Engineering, Yanshan University, Qinhuangdao, China. Her current research interests include robust adaptive control of nonlinear systems and time-delay systems and applications to hybrid distributed generation systems and automotive powertrain.

Mingxin Kang earned his PhD in mechanical engineering from Sophia University, Tokyo, Japan, in 2015. For the past three years, he applied himself to the study of engine transient control and real-time optimization control theory. His main contributions include the transient control scheme development for the engine-in-the-loop simulation system and nonlinear receding horizon optimal control for the gasoline engine control system. He currently serves as a postdoctoral research fellow in Sophia University. His current research interests include automotive energy optimization and nonlinear optimal control for the engine system.

Junichi Kako received a BE degree from Nagoya Institute of Technology, Nagoya, Japan. He joined Toyota Motor Corporation, Japan, in 1989. He worked on various aspects of automotive powertrain control. From 1989 to 1994, he was part of the team involved in the development of laboratory automation system, engineering office automation system, and embedded system of powertrain control. From 1995–2001, he focused on the engine control systems in the Powertrain Management Engineering Division. In 2002, he was with the Future Project Division wherein he was responsible for the R&D of model-based engine control system. Currently, he is involved with developing engine control systems in the Advanced Engine Management System Development Division, Toyota Motor Corporation.

Akira Ohata received a Master's Degree from Tokyo Institute of Technology, Tokyo, in 1973, and directly joined Toyota Motor Corporation, Shizuoka. He was involved in exhaust gas emission controls, development of intake and exhaust system, including variable intake systems, hybrid vehicle control system, vehicle controls, model-based development, and the education of advanced control theory at Toyota. He is responsible for the standardization activity in object management group assuring dependability of consumer devices. His current major interest is modeling which includes model simplification and the integration of physical and empirical models. He is a senior general manager at Toyota Motor Corporation, a vice chair of IFAC TC7.1 (automotive control), a research fellow of Information Technology Agency (IPA) under the Ministry of Economy, Trade, and Industry, and the chair of technical committee on plant modeling of SICE (Society of Instrument and Control Engineers). He received an award for the most outstanding paper in Convergence 2004 and an award for technical contribution from Japan Society of Automotive Engineering (JSAE).

1

Introduction

In recent years, improving the efficiency of automotive powertrains has attracted much attention since, among total energy consumption in the world, transportation energy use has amounted to more than 26%, and as reported by [1], road vehicles account for three-quarters of the total transportation energy (Figure 1.1). Moreover, 95% of the transportation energy comes from oil-based fuels, largely diesels and gasoline. Although a lot of alternative vehicles such as hybrid electric vehicles and plug-in hybrid electric vehicles have been marketed, more than 85% of road vehicles still load combustion engines on the powertrains. As predicted in [2], this situation will be continued, and more than 85% of road vehicles will load combustion engines within the coming 30 years [2].

Meanwhile, in the past two decades, rapid progress in car electronics and digital processing technology has changed the combustion engines into computerized machines. This trend enables us to improve the engine efficiency by real-time control technology. However, engine management in the past, even if implemented by an electronic control unit (ECU, see Figure 1.2) was mainly focused on the static or pseudostatic mode, since handling the behavior of combustion engines at the transient mode is very difficult due to the strong nonlinearities and uncertainties in the thermodynamics, fluid dynamics in the inlet and exhaust paths and fuel paths, and so forth.

Moreover, the combustion engine is an old topic in thermal engineering where attention was focused on stable combustion and efficiency improvement evaluated in the sense of cyclic average. Dealing with the combustion engine as a dynamical system and approaching the goal of efficiency improvement with advanced control theory have recently been spotlighted; therefore, engine control becomes a very active field in control engineering.

Indeed, much literature has been published in the last decade that focused on topics such as powertrain systems, driveline systems, combustion engines, and the vehicles [3, 4]. However, regarding combustion engines from the view of dynamical system control is still a fruitful topic. The purpose of this book is to present the fundamentals of modeling and control design for engine transient operation. The presented approaches are model-based control methods to handle the transient dynamics of the combustion engines based on the control-oriented models.

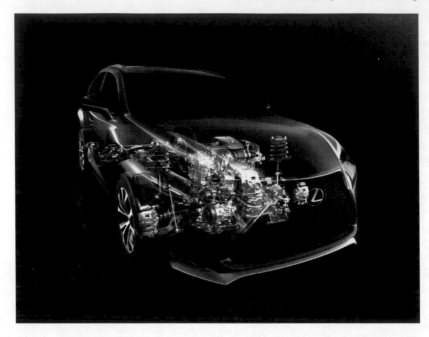

FIGURE 1.1
Automotive powertrains account for three-quarters of the total transportation energy.

FIGURE 1.2
Combustion engine and ECU.

1.1 Control-Oriented Engine Model

In internal combustion engines, the dynamical characteristics are determined by many factors, such as fluid dynamics of the air charge and fuel injection paths, in-cylinder thermal dynamics, and mechanical dynamics. Furthermore, the combustion efficiency is also influenced by the chemical and physical properties of the fuel and the environmental conditions, which have stochastic characteristics. Therefore, modeling the engine dynamics based on a physics-based model will lead to complicated mathematical representation with high dimension and strong nonlinearity, uncertainty, and probability.

At the early stage, static mapping is used in engine control to adjust the actuating signals according to the engine operating conditions. To describe the transient response of the engine speed under the throttle operation, the first-order transfer function with delay, which represents the intake-to-power delay, as sketched in Figure 1.3, is exploited in the engine modeling field. Obviously, this is a speed control-oriented model, and only the input–output behavior from the throttle opening to the speed response is represented in the sense of mathematical equality.

It should be noted that the transfer function works only for the linear and time invariant determinate systems, and no more information of the internal states of the system is provided by the transfer function. To handle the nonlinearity and the internal behavior of combustion engines without increasing the model complicity, the so-called mean-value model was proposed [5, 6] in the engine control community. This modeling approach focuses on the internal

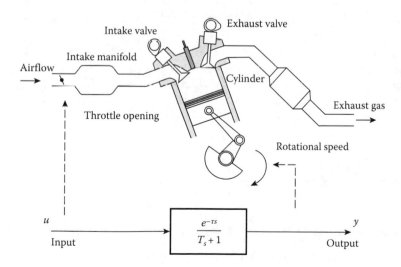

FIGURE 1.3
Engine dynamics and the first-order model.

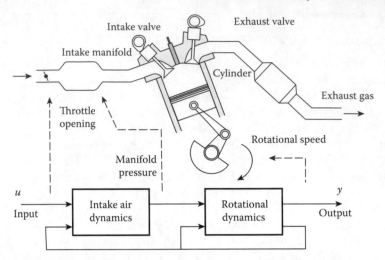

FIGURE 1.4
Mean-value model.

states of the engine and deals with them in the sense of cyclic or fixed-horizon mean-value characteristics. For example, focusing on the cyclic air charge mass, or the torque generated by combustion, and calibrating it with the engine speed and the intake manifold pressure, a second-order nonlinear differential equation is obtained in [5] to represent the dynamical behavior from the throttle opening to the speed response with the information of the internal states. As shown in Figure 1.4, this kind of mean-value model consists of interconnected subsystems and provides the response of the internal states.

Very recently, model-based development (MBD) has been spotlighted in the automotive industry in pursuit of efficiency of product development. This trend requires the engine models to be able to provide the representation of in-cylinder behavior, including the heat-release profile and the cylinder-to-cylinder or cycle-to-cycle differences. Of course, this kind of in-cylinder model increases the complicity of the model structure and the computing load.

Reviewing the history of the engine model, it is always control purpose oriented and influenced by the progress of digital computing technology. Figure 1.5 shows the trend of engine models. In this book, the engine model exploited in each chapter is not a unified one, which is reconstructed to match the control objectives discussed in each chapter.

1.2 Control Issues of Gasoline Engine

For combustion engines with four strokes, management of the air charge mass and the fuel filling mass per cycle is a most important subject for the

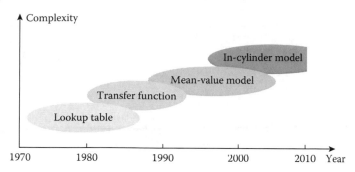

FIGURE 1.5
Trend of engine model development.

transient control, since the combustion performance and the torque generation are greatly affected by the ratio of the two masses and the total amount. Furthermore, there are a lot of factors that influence the combustion efficiency, such as the spark timing, the intake and the exhaust valve timing, and the thermal conditions in the cylinders. Roughly speaking, the challenges in engine control, especially at the transient mode, are how to exactly handle the air charge and the fuel mass before deciding the actuation, such as throttle opening, fuel injection, and spark advance (SA).

As is well known, when the engine is operating at a static mode, that is, the engine speed and the load kept invariant, the average mass flow rates of the inlet air and the filling fuel are equivalent to the gas mass flow rate exhausted from the tailpipe. Therefore, with the oxygen sensor equipped at the exhaust manifold, it is not difficult to handle the air charge and fuel filling mass. However, at the transient operation, handling the mass with this idea will lead to unacceptable error. For instance, in the air–fuel ratio (A/F) control, deciding the fuel injection mass with the average air charge mass calibrated at the corresponding static operation mode cannot satisfy the A/F control performance. Figure 1.6 shows experimental results conducted on a gasoline engine where the A/F is disrupted under fuel cutting and speed changes, respectively. In order to improve the performance, the fuel injection mass must be decided with some predictive mechanism, and to this end, a mathematical model representing the transient dynamics is usually exploited in real-time engine control.

In this book, the following control issues will be addressed in the fashion of control-oriented model-based control.

Speed Control. Speed control is an old topic in engine control. Usually, the throttle opening is chosen as the primary control actuator and the SA is as the secondary actuator to reject the quick disturbances in the speed response. Focused on the transient mode, such as starting engine and acceleration, the speed tracking control problem will be addressed with nonlinear tracking control and receding horizon optimal control approaches, respectively. In the

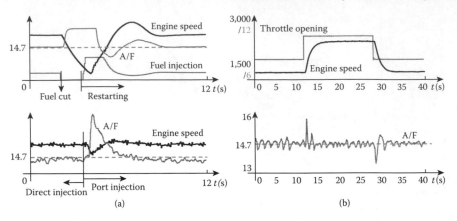

FIGURE 1.6
Transient responses of gasoline engines. (a) Transient behavior of changing fuel path. (b) Transient behavior of changing throttle valve.

latter, only the dynamics from the throttle opening to the speed is considered, along with the mean-value model concerning the intake air dynamics and the mechanical dynamics of the crankshaft.

A/F Control. A/F control is a fundamental issue in internal combustion engines. The challenge in A/F control is how to handle the air charge mass, especially during transient operation mode. Moreover, the parameters of the air intake path and the fuel path are usually difficult to determine with certainty. A Lyapunov-based design to achieve a nonlinear feedback control scheme is derived based on the mean-value model where the Lyapunov stability theorem is used to guarantee the boundedness of the control parameter and the convergency of the A/F control error.

Receding Horizon Optimal Control. As a real-time optimization strategy, receding horizon optimal control makes the most of the transient model for performing optimization of engine performance with real-time states of the engine. Hence, it is also called model predictive control (MPC) and has recently attracted much attention. The bottleneck in applying this kind of control algorithm to combustion engines is the on-line computing loads. The so-called continuation and generalized minimum residual estimation (continuous/GMRES) algorithm is exploited in this book, and based on the algorithm, the speed and torque control problems are solved with the same design framework.

Balancing Control. In multicylinder engines, cylinder-to-cylinder or cycle-to-cycle imbalance is due to the stochastic characteristics of the combustion event. However, many feedback control loops are performed based on one sensor that usually provides the mean value of the targeted control output. For example, the A/F control is usually conducted based on the oxygen sensor equipped at the gas mixing point of the exhaust manifold, which provides the

A/F of the mixing gas exhausted from multicylinders. Hence, to decide the fuel injection mass cylinder by cylinder is challenging under the constraint of using only one sensor. Cylinder-to-cylinder A/F ratio control will be addressed in this book with the autoregressive and moving-average (ARMA) model and recursive identification algorithm.

Residual Gas Fraction. Residual gas is the gas burned during previous cycles that was left in the cylinder. In order to use the residual gas to increase the heat capacity of the air charge and improve the engine performance, management of the residual gas fraction (RGF) with respect to the total air charge is an important issue to improve the combustion quality. However, due to the stochastic characteristics of combustion, it is difficult to handle the RGF via the actuators that effect the residual gas mass. A stochastic control approach to the RGF is introduced to model the behavior of RGF, and control is in the sense of mean value.

In Chapter 8, three benchmark problems are explained that are proposed by the authors as an academic–industrial collaboration. The benchmark problems will provide a challenging opportunity for the readers to be initiated into the modeling and control of combustion engines. Furthermore, the basic concepts with necessary fundamentals are briefly summarized in appendixes, so that this book becomes a self-contained document for readers who are unfamiliar with control theory.

1.3 Experimental Setup

As an engineering book jointly written by the authors from academic and industry comity, this book is concerned not only with design theory, but also with experimental practices. All experimental results introduced in the following chapters are conducted on a full-scale gasoline engine test bench. A brief overview of the research facility is as follows.

In the engine test system, the dynamometer frontier is coupled to the engine crankshaft, and it can adjust the engine operating conditions by setting different work modes for the dynamometer. In general, there are two kinds of working modes to be adopted in the testing process: speed control mode and torque control mode. In the former mode, the actual dynamometer speed is controlled to follow a given speed command, and actually this action will further induce the change of the engine speed; in the latter mode, the dynamometer torque output is treated as the control variable, and it will be controlled to track the reference torque command. With such a configuration, the engine speed or output torque can be well adjusted and calibrated under steady operating conditions according to the experimental requirements because of the coupling relations between the engine and dynamometer. However, one key point for the transient engine control test is to reduce the transient response time of the dynamometer. To this end, the electrical

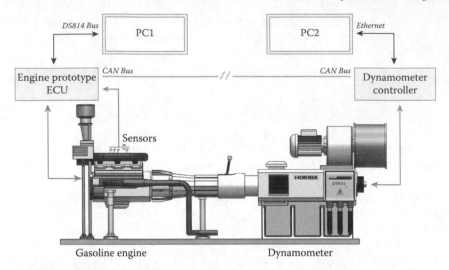

FIGURE 1.7
General structure of the engine-dynamometer test bench.

FIGURE 1.8
Testing facilities and control platform of the test bench. (a) Engine-dynamometer test bed. (b) System control platform.

dynamometer with low rotational inertia becomes a significant option for the engine transient tests. A state-of-the-art engine-dynamometer test bench was constructed in the Advanced Powertrain Control Laboratory of Sophia University, and the general structure of the test bench is shown in Figure 1.7. The actual test environments and control platform are shown in Figure 1.8.

In the test bench, a 3.5 L V6-type full-scale gasoline engine is adopted to connect with an alternating current (AC) electrical dynamometer. The specification of the gasoline engine is shown in Table 1.1. In order to achieve real-time control for the gasoline engine, an engine prototype ECU is equipped, in which a rapid prototype controller (dSPACE) is exploited as a parallel develop tool

TABLE 1.1
Gasoline engine specification

Type	2GR-FSE (Toyota, Inc.)
No. of cylinders and arrangement	6 cylinders, V-type
Manifolds	Parallel-flow
Fuel system	Port and direct injection
Displacement	3,456 cm^3
Max. output	228 kW @ 6,400 rpm
Max. torque	375 Nm @ 4,800 rpm

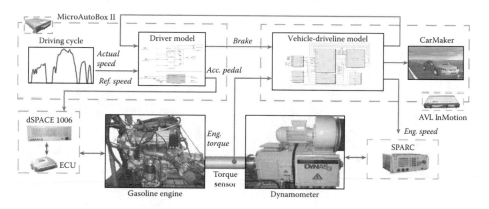

FIGURE 1.9
Block diagram of engine-in-the-loop simulation system.

and connected with the engine stock ECU via the bypass interface. Herein, our control scheme can be easily programmed in the Simulink® platform, and then it can be compiled and executed in real time in this parallel controller. With such a configuration, the dSPACE controller can take over any control loop, such as throttle control, fuel injection control, SA control, and variable valve timing (VVT) control, by overwriting the ECU's commands with external control commands. Meanwhile, to monitor the engine operating conditions, some auxiliary sensors are installed, including an intake manifold pressure sensor, encoder, lambda sensor, in-cylinder pressure sensor, and exhaust temperature sensor. All sensor signals can be collected by the dSPACE controller in real time. On the dynamometer side, the real-time control of the low-inertia dynamometer can be achieved by an advanced controller (SPARC), which indeed can perform a satisfactory transient control performance.

In practice, the above test configuration provides an efficient and repeatable experimental platform to evaluate the controller design and calibrate engine transient performance. Moreover, to simulate the operating environment for the engine under any driving conditions, an engine-in-the-loop simulation (EILS) system is developed based on the proposed engine-dynamometer test bench. The structure of the EILS is sketched in Figure 1.9.

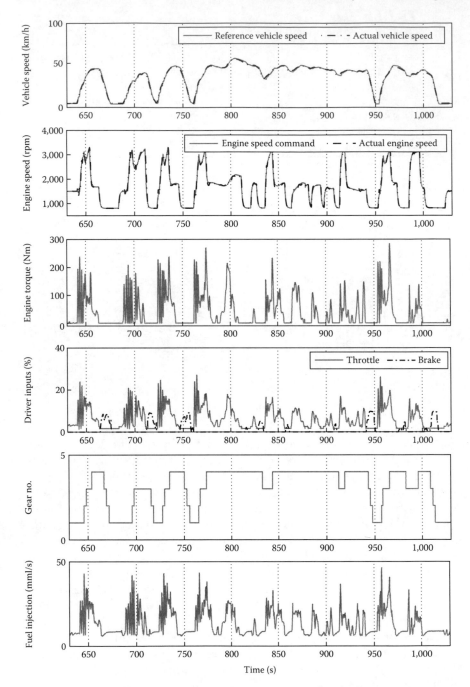

FIGURE 1.10
EILS experimental result with driving cycle.

In the system, the real engine-dynamometer test bench is coupled with a virtual vehicle mathematical model. The dynamometer is treated as an external load simulator, and it adjusts the engine operating conditions in real time. The main concept of such a system is to guarantee that the virtual vehicle can be driven by the real engine under any driving conditions. In fact, the virtual system consists of two independent simulation modules. One is the driver control module, which can be programmed in Simulink and implemented in real time on the MicroAutoBox II. Besides, another virtual module is the vehicle–driveline model. A commercialized simulation software referred to as CarMaker is adopted here for providing the accurate vehicle dynamic model as well as the real-time driving scenario. In addition, a hardware named InMotion provides an interface platform for the vehicle–driveline model with other system components. Indeed, the data exchange among these different hardware equipments is achieved via the controller area network (CAN) bus technique.

In the real-time implementation, when the route information and reference vehicle speed are given, the driver model will identify the operation commands, such as acceleration, brake, and gearshift, and then transmit the relevant commands to the vehicle model and engine prototype controller, respectively. Accordingly, the engine will output the torque. And this torque will be measured by the torque sensor and drag the virtual vehicle forward to follow the reference vehicle speed. At the same time, a reference engine speed will be estimated by the driveline model based on the instant vehicle speed and the gear ratio, and it is fed back to the dynamometer controller in real time. Note that the dynamometer works in the speed control model, and consequently, it will drag the engine to follow the given reference speed command with a very fast response time. In this closed-loop control system, the transient control performance of the dynamometer is crucial to the simulation accuracy and operation safety. A partial testing result based on the proposed simulation system is illustrated in Figure 1.10.

It should be pointed out that the proposed EILS system realizes the situation in which the driver drives the vehicle with a real engine on the road. Besides, owing to the repeatability of test conditions, the system can be used not only for the transient control validation, but also for the optimization of fuel efficiency with consideration of the traffic environment.

2

Mathematical Model of Gasoline Engines

2.1 Introduction

Figure 2.1 shows a schematic representation of a spark ignited (SI) engine. For four-stroke engines that consist of the intake stroke, compression stroke, combustion stroke, and exhaust stroke, thermal energy is generated by the combustion of the compressed air–fuel mixture, and the thermal energy is transferred as the force acting on the piston. The force depends on the efficiency of the released energy conversion, which is called heat release, the air charge during the intake stroke, and the fuel injection. When the combustion finishes, the gas is pushed out through the exhaust valve. The opening–closing timing of the intake valve determines the amount of air charge, and the intake air mass flow rate depends also on the pressure and temperature of the intake manifold and the pressure and temperature of the cylinder. Hence, the variations of the pressure and temperature in the intake manifold cause the variation of the intake air mass flow rate. Furthermore, the produced thermal energy and the variations of the air mass flow rate passing the throttle valve and the intake valve follow the mass and energy conservation laws.

After two rotational revolutions of the crankshaft in this way, one cycle of the four-stroke engine is completed, and the next cycle follows. In other words, during one cycle, the engine torque is generated only once during the combustion, that is, during the 180 degrees in crank angle. For the engines with multicylinders, the cylinders are displaced according to a mechanical phase rule, and then the combustion stroke of each cylinder follows serially, and the force acting on the crankshaft is serially generated. Therefore, the rotational speed of the crankshaft is influenced by the amount of the air–fuel mixture, which is determined by the actuators, including the opening of the throttle valve, the individual opening–closing timing of the intake valve, the fuel injection, and the spark timing.

The above description shows that engine dynamics is determined by the physical principles in terms of mechanics, fluid, and combustion chemistry, such as Newton's law and the first law of thermodynamics. However, it is not easy to precisely characterize the relevant phenomena. Moreover, from the perspective of system control, the engine exhibits a hybrid system that includes continuous and discrete dynamics. In this chapter, engine dynamics with modeling issues is introduced. The fundamentals concerning the rotational motion

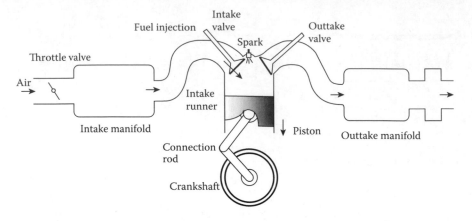

FIGURE 2.1
Schematic of engine cylinder system.

of the engine are introduced first. Then, a mathematical model that represents the dynamics of the internal combustion engine is addressed.

2.2 Physics

For the engine with multicylinders, the generated torque of the individual cylinder is mainly determined by the mass of the air charge and the fuel injection; moreover, it is also affected by the spark timing, the residual gas, the combustion efficiency, and so on. In fact, the fresh air enters the intake manifold through the throttle valve and is inducted into each cylinder through the intake valve of the individual intake port. This route is often called the air intake path. By applying certain pressure to the injectors, which are located at each intake port, the fuel is injected by the injectors and enters each cylinder with the inducted air. This process is called the fuel path. After a compression stroke in each cylinder, there is a combustion stroke, that is, the combustion process of the inducted air–fuel mixture. The pressure produced by the combustion acts on the piston to drive the connection rod for the rotational operation of the crankshaft. The above description is called the torque generation process. The generated torque in individual cylinders drives the rotational motion of the crankshaft.

Therefore, considering the throttle opening, fuel injection, and spark timing as actuators and dealing with engine speed as output, the air intake and fuel paths, the torque generation, and the mechanical rotational motion are the main components of the engine dynamical system, as shown in Figure 2.2. Moreover, the dynamics of each component depends on the crank angle, and this fact makes it easier to understand their effects to the rotational speed.

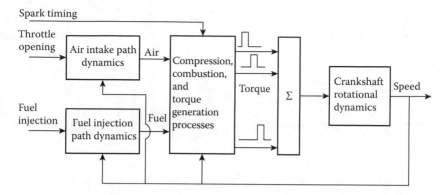

FIGURE 2.2
Structure of engine dynamics.

The detailed engine model can be deduced based on the analysis of the characteristics of each component. Fundamental rules with respect to the physics are essential to achieve the modeling. For the air with fixed volume, such as in the intake manifold and the cylinder during the compression stroke, the ideal gas law is used to describe the variations of the air states. The mass conservation law and the first law of thermodynamics are useful for dealing with the energy transfer issue. Furthermore, the basic principle for the flow of fluids through an orifice is important for modeling the pipes with valves. Finally, the mechanical motion of the crankshaft is determined by Newton's law.

As a reference, these essential physical laws are introduced in detail in the following.

Ideal Gas Law. The states of a hypothetical amount of ideal gas with mass m (kg) satisfy the following ideal gas state equation:

$$pV = mRT \tag{2.1}$$

where p (Pa), V (m^3), and T (K) denote the states of the gas, that is, the pressure, volume, and temperature, respectively, and R (J/[kg·K]) denotes the ideal gas constant.

First Law of Thermodynamics. The first law of thermodynamics is the conservation of energy principle specialized for thermodynamical systems. For the gas with a fixed volume, the absorbed heat from outside of the system (dq), the work done by the gas to the outside (dw), and the variation in internal energy of the gas (du) satisfy the following energy conservation law:

$$du = dq - dw \tag{2.2}$$

Specific heat capacity is a physical quantity for measuring the energy changes due to the changes of the gas states. The specific heat c_v (J/[kg · K]) represents the ratio of the absorbed thermal energy to the variation in temperature of unit mass of the gas at constant volume, and correspondingly, the specific heat c_p (J/[kg · K]) represents the ratio of the absorbed thermal energy to the variation in temperature of unit mass of the gas at constant pressure. The ratio of the specific heats is denoted as $\kappa = c_p/c_v$. Using the specific heats, the variation of the internal energy of the gas at a constant volume process can be described as follows.

First, since no work is done by the gas at constant process, $dw = 0$. From the first law of thermodynamic, the following relationship is obtained:

$$du = dq \tag{2.3}$$

Then for unit mass of gas, the absorbed heat can be expressed as $dq = c_v dT$; hence, due to (2.3), the internal energy change of the gas

$$du = c_v m dT \tag{2.4}$$

can be calculated. Furthermore, discussions in the next section will depend on the following relationships on the specific heats. The total energy of a gas system is measured by the enthalpy h (J), which is defined as

$$h = u + pV \tag{2.5}$$

where u denotes the internal energy. At a constant pressure process, the work done by the gas is proportional to the variation of the volume. In other words, $dw = pdV$. Note that $dp = 0$; then the variation in the enthalpy is $dh = du + pdV$, which from (2.2) can be calculated as

$$dh = du + pdV = du + dw = dq = c_p dT \tag{2.6}$$

Generally, values of the specific heats c_p and c_v are dependent on the temperature T. For ideal gas, c_p and c_v are constant, and from the enthalpy (2.5) and the ideal gas state equation (2.1), the following expressions can be obtained:

$$c_p = \frac{\kappa}{\kappa - 1} R, \quad c_v = \frac{1}{\kappa - 1} R, \quad c_p - c_v = R \tag{2.7}$$

Airflow Passing through an Orifice. Consider the airflow passing through an orifice as shown in Figure 2.3. The air mass flow rate can be calculated by the following equation [7]:

$$\dot{m} = \begin{cases} \dfrac{Ap_0}{\sqrt{RT_0}} \left(\dfrac{p}{p_0}\right)^{\frac{1}{\kappa}} \sqrt{\dfrac{2\kappa}{\kappa - 1}\left[1 - \left(\dfrac{p}{p_0}\right)^{\frac{\kappa-1}{\kappa}}\right]}, & \dfrac{p}{p_0} \geq \left(\dfrac{2}{\kappa+1}\right)^{\frac{\kappa}{\kappa-1}} \\[4mm] \dfrac{Ap_0}{\sqrt{RT_0}} \sqrt{\kappa \left(\dfrac{2}{\kappa+1}\right)^{\frac{\kappa+1}{\kappa-1}}}, & \dfrac{p}{p_0} < \left(\dfrac{2}{\kappa+1}\right)^{\frac{\kappa}{\kappa-1}} \end{cases} \tag{2.8}$$

where m (kg) denotes the air mass, A (m^2) denotes the cross section of the orifice, p_0 (Pa) and T_0 (K) denote the upstream pressure and temperature, and p denotes the downstream pressure.

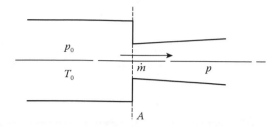

FIGURE 2.3
Airflow passing through an orifice.

The model (2.8) is deduced based on the engine conversation principle. More details are shown in Section 2.8.

2.3   Modeling of Components

Descriptions for the modeling of the air intake path and fuel path are given in this section.

2.3.1   Air Intake Path

Regarding the air intake path, assume that the fluid of each component is an ideal gas with constant specific heats c_p and c_v. Moreover, let p_a and T_a denote the air pressure and temperature, and p_m, T_m, V_m, and m denote the pressure, temperature, volume, and air mass of the intake manifold.

Figure 2.4 shows a schematic overview of the air intake path of a four-cylinder engine. In the following, the engine with n cylinders is considered, and for the case of $n = 4$, the conclusion is the same. First, to model the air intake path dynamics, the air in the intake manifold and each intake pipe is considered a lumped parameter system. The air mass flows passing through the throttle valve opening to each cylinder through each intake pipe (called intake port), and the intake valves are characterized by using the model of the airflow passing through an orifice presented in Section 2.2. Furthermore, assuming that the fluids are continuous compressible one-dimensional isentropic flows, the dynamical equations with respect to each volume airflow are deduced based on the ideal gas law as well as the energy conservation law.

State Equation of Intake Manifold. As shown in Figure 2.4, consider the situation that the intake pipes of the n cylinders share one intake manifold. Based on the mass conservation law, the changes of the air mass in the intake manifold equal the difference between the air mass flow rate passing through

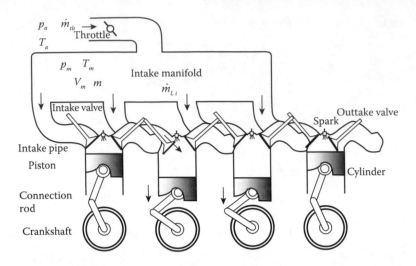

FIGURE 2.4
Schematic of air intake path of a four-cylinder engine.

the throttle into the manifold $\dot{m}_{th}$ and the air mass flow rate flowing into each intake pipe $\dot{m}_o$:

$$\dot{m}(t) = \dot{m}_{th}(t) - \dot{m}_o \tag{2.9}$$

Moreover, let $\dot{m}_{l,i}$ denote the air mass flow rate flowing into the intake pipe of cylinder i; then $\dot{m}_o = \sum_{i=1}^{n} \dot{m}_{l,i}(t)$.

Furthermore, let dq_m denote the total heat variation, dQ_m the absorbed heat, du_m the change of the internal energy, and dw_m the corresponding mechanical work done by the air. Note that the manifold volume is constant, which means that there is no mechanical work done by the air in the manifold, that is, $dw_m = 0$. Moreover, denote the heat transfer within time dt from the environment into the air in the manifold as $\dot{Q}_m dt$.

The air mass $\dot{m}_{th}dt$ passing through the throttle valve into the manifold also brings heat into the manifold, and the air mass $\dot{m}_o dt$ going into each intake pipe takes heat away from the air in the manifold, where the corresponding heat changes are $c_p T_a \dot{m}_{th} dt$ and $c_p T_m \dot{m}_o dt$, respectively. Then, the total heat transfer into the air system in manifold can be calculated as follows:

$$dq_m = \dot{Q}_m dt + c_p T_a \dot{m}_{th} dt - c_p T_m \dot{m}_o dt \tag{2.10}$$

Finally, note that the change of the internal energy of the air system in the manifold satisfies $du_m = d(c_v m T_m)$ then using the first law of thermodynamics (2.2) gives

$$\frac{d(c_v m T_m)}{dt} = \dot{Q}_m + c_p T_a \dot{m}_{th} - c_p T_m \dot{m}_o \tag{2.11}$$

Taking into account the ideal gas state equation $mT_m = p_m V_m / R$ and the relationships of the specific heats (2.7), we obtain the following equation from (2.11):

$$\frac{d}{dt}\left(\frac{p_m V_m}{\kappa - 1}\right) = \dot{Q}_m + \frac{\kappa R}{\kappa - 1}(T_a \dot{m}_{th} - T_m \dot{m}_o)$$

Assuming that κ is constant, the above equation can be rearranged as

$$\dot{p}_m = \frac{\kappa - 1}{V_m}\dot{Q}_m + \frac{\kappa R}{V_m}\left(T_a \dot{m}_{th} - T_m \sum_{i=1}^{n} \dot{m}_{l,i}\right) \tag{2.12}$$

On the other hand, substituting the ideal gas state equation into the mass conservation equation (2.9) yields

$$\frac{d}{dt}\left(\frac{p_m V_m}{R T_m}\right) = \dot{m}_{th} - \dot{m}_o$$

which can be rearranged as

$$\frac{dT_m}{dt} = \frac{T_m}{p_m}\frac{dp_m}{dt} - \frac{R}{V_m}\frac{T_m^2}{p_m}\left(\dot{m}_{th} - \sum_{i=0}^{n} \dot{m}_{l,i}\right)$$

Submitting Equation 2.12 for dp_m/dt in the first term of the above equation obtains

$$\dot{T}_m = \frac{\kappa - 1}{mR}\dot{Q}_m + \frac{1}{m}(\kappa T_a - T_m)\dot{m}_{th} - (\kappa - 1)\frac{T_m}{m}\sum_{i=1}^{n} \dot{m}_{l,i} \tag{2.13}$$

Therefore, the air state equation of the manifold is represented by (2.9), (2.12), and (2.13). Moreover, the air mass flow rate passing through the throttle valve $\dot{m}_{th}$ and the air mass flow rate flowing into each intake port $\dot{m}_{l,i}$ are determined by the fluid equation for the airflow passing through an orifice, that is,

$$\dot{m}_{th} = A_{th}(\phi)\psi(p_m, p_a, T_a) \tag{2.14}$$

$$\dot{m}_{l,i} = \begin{cases} A_i \psi(p_{r,i}, p_m, T_m), & p_m \geq p_{r,i} \\ A_i \psi(p_m, p_{r,i}, T_{r,i}), & p_m < p_{r,i} \end{cases} \tag{2.15}$$

where A_i denotes the cross section of the ith intake port and $A_{th}(\phi)$ denotes the active area of the throttle valve, which can be calculated by

$$A_{th}(\phi) = \pi r^2 (1 - \cos \phi) \tag{2.16}$$

with the radius of the throttle valve r and the throttle opening ϕ, and $\psi(p, p_0, T_0)$ $(p_0 \geq p)$ given by the following nonlinear function:

$$
\psi(p, p_0, T_0) = \begin{cases} \dfrac{p_0}{\sqrt{RT_0}} \left(\dfrac{p}{p_0}\right)^{\frac{1}{\kappa}} \sqrt{\dfrac{2\kappa}{\kappa - 1} \left[1 - \left(\dfrac{p}{p_0}\right)^{\frac{\kappa-1}{\kappa}}\right]}, & \dfrac{p}{p_0} \geq \left(\dfrac{2}{\kappa+1}\right)^{\frac{\kappa}{\kappa-1}} \\[2em] \dfrac{p_0}{\sqrt{RT_0}} \sqrt{\kappa \left(\dfrac{2}{\kappa+1}\right)^{\frac{\kappa+1}{\kappa-1}}}, & \dfrac{p}{p_0} < \left(\dfrac{2}{\kappa+1}\right)^{\frac{\kappa}{\kappa-1}} \end{cases}
$$

$$(2.17)$$

Intake Port. Notice that each intake port is with a constant volume from its inlet port to its intake valve; hence, the state equation of the air in each intake port can be deduced by the same discussion as the air system of the intake manifold. In this case, let $\dot{m}_{l,i}$ denote the air mass flow rate flowing into the intake port of cylinder i and $\dot{m}_{c,i}$ denote the air mass flow rate passing through the intake valve into the cylinder. Then, a same derivation for Equations 2.9, 2.12, and 2.13 obtains the following state equations (details are omitted here):

$$
\begin{cases} \dot{p}_{r,i} = \dfrac{\kappa - 1}{V_{r,i}} \dot{Q}_{r,i} + \dfrac{\kappa R}{V_{r,i}} \left(T_m \dot{m}_{l,i} - T_{r,i} \dot{m}_{c,i}\right) \\[1.2em] \dot{T}_{r,i} = \dfrac{1}{m_{r,i}} \dot{Q}_{r,i} + \dfrac{\kappa T_m - T_{r,i}}{m_{r,i}} \dot{m}_{l,i} - (\kappa - 1)\dfrac{T_{r,i}}{m_{r,i}} \dot{m}_{c,i} \\[1.2em] \dot{m}_{r,i} = \dot{m}_{l,i} - \dot{m}_{c,i} \end{cases} \qquad (2.18)
$$

where $p_{r,i}$, $T_{r,i}$, and $V_{r,i}$ denote the air pressure, temperature, and volume of the intake port of cylinder i, and $\dot{Q}_{r,i}$ denotes the heat transfer into the air system of the intake port of cylinder i due to the changes of the environment. In addition, considering the air pressures upstream and downstream with respect to each intake valve, the air mass flow rate $\dot{m}_{c,i}$ can be represented as

$$
\dot{m}_{c,i} = \begin{cases} A_{v,i}(L_{v,i}) \psi(p_{c,i}, p_{r,i}, T_{r,i}), & p_{r,i} \geq p_{c,i} \\[0.8em] A_{v,i}(L_{v,i}) \psi(p_{r,i}, p_{c,i}, T_{c,i}), & p_{r,i} < p_{c,i} \end{cases} \qquad (2.19)
$$

where $A_{v,i}$ denotes the active area of the intake valve of cylinder i, which depends on the valve lift $L_{v,i}$.

In general, the volume of the intake port is quite small compared to the manifold volume. Hence, the air dynamics of the runners is neglected, and the air mass flow rate passing through the intake port of each cylinder is usually considered equal to the air mass flow passing through the intake valve. In this case, $\dot{m}_{l,i} = \dot{m}_{c,i}$ $(i = 1, 2, \cdots, n)$.

2.3.2   Fuel Path

In port-injected gasoline engines, the fuel is injected intermittently by the injector, which is usually located near the intake valve of each intake port.

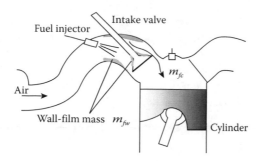

FIGURE 2.5
Fuel injection model.

However, the injected fuel in the current cycle partially enters into the cylinder. Part of the fuel is absorbed at the back of the intake valve and at the intake port wall, and this part of the fuel can also enter the cylinder by vaporization along with the intake air. Exact modeling of this process is quite complex.

The fuel path dynamics is generally represented by using the wall-wetting model that represents the behavior from the fuel injection command to the fuel mass injected into the cylinder with a first-order differential equation [13]

$$\begin{cases} \dot{m}_{fw} = -\tau_w m_{fw} + \varepsilon_w u_f \\ m_{fc} = \tau_w m_{fw} + (1 - \varepsilon_w) u_f \end{cases} \tag{2.20}$$

In the model (2.20), u_f denotes the injected fuel mass command, m_{fc} denotes the corresponding fuel mass injected into the cylinder, m_{fw} is the state variable of the dynamics, which denotes the total absorbed fuel mass as shown in Figure 2.5, and τ_w and ε_w are the model parameters that are dependent on the engine shape, engine speed, external load, and so forth.

For the engines with a direct injection system, the fuel is delivered to the cylinder directly, and its mass is identical to the injected fuel mass command, that is,

$$m_{fc} = u_f \tag{2.21}$$

2.4 In-Cylinder Dynamical Model

The in-cylinder state is transited cycle by cycle, and each state is with the intake stroke where the air and fuel are brought into the cylinder. The physics of the intake stroke is as mentioned in Section 2.3. Then, following the intake stroke, the air–fuel mixture entering the cylinder through the air intake path and the fuel path is compressed during the compression stroke. During this stroke, the variations of the states in the manifold and intake port do not influence the charged air mass and the fuel mass of the cycle. After the compression, the air–fuel mixture is ignited to start the combustion stroke. The thermal energy due to combustion is transferred to the force acting on the piston; in other words, the torque is generated to drive the rotational movement of the crankshaft. Moreover, from the viewpoint of signal transmission, there exist time delays from the pressure signals of the manifold and intake port to the torque generation. This time delay (denoted as t_d) is associated with the crank angle and can be approximated by π (rad), that is, $t_d = 30/n_e$ s, where n_e (rpm) denotes the engine speed.

In the following, modeling the in-cylinder air dynamics is deduced based on the energy conservation law. The basic principles are the same as the modeling of the air behavior in the manifold and the intake port. In this situation, the energy transfer to the environment with respect to the states during the opening and closing of the intake and outtake valves is considered. However, the heat transfer between the air and the cylinder wall is neglected during the whole cycle.

In the following, for cylinder i, let $\theta_{c,i}$, $\theta_{s,i}$, $\theta_{b,i}$, and $\theta_{e,i}$ denote the crank angle at the time when the intake valve closes, at the spark timing, at the time when combustion is completed, and at the time when the outtake valve opens, respectively. These moments with respect to each discrete event satisfy the condition $0° < \theta_{c,i} < \theta_{s,i} < \theta_{b,i} < \theta_{e,i} < 720°$. Moreover, define the phase angle of cylinder i, $\theta_i = 0$, when the piston is at the top dead center (TDC) with respect to the intake stroke. Then, θ_i with the crank angle θ has the following relationship:

$$\theta_i = \mathbf{mod}(\theta - (i-1)180, 720), \qquad i = 1, 2, \cdots, n \qquad (2.22)$$

Furthermore, let $p_{c,i}$, $T_{c,i}$, and $V_{c,i}$ denote the pressure, temperature, and volume and mass of the air, respectively.

First, unlike the air in the manifold, there is work done by the in-cylinder air to the environment due to the reciprocating motion of the piston. Due to pressing the piston, the variation of the mechanical work done by the air can be formulated as $dw = p_{c,i}dV_{c,i}$, and taking the time derivative yields

$$\dot{w} = p_{c,i}\dot{V}_{c,i} \qquad (2.23)$$

where the variation of $V_{c,i}$ is dependent on the cylinder volume with respect to the crank angle.

Then, consider the variation of du is given by the mass and temperature of the in-cylinder air:

$$du = d(c_v m_{c,i} T_{c,i}) = d\left(\frac{c_v}{R} p_{c,i} V_{c,i}\right) \qquad (2.24)$$

Assuming that c_v and R are constant, obtain the following differential equation:

$$\dot{u} = \frac{c_v}{R}(\dot{p}_{c,i}V_{c,i} + p_{c,i}\dot{V}_{c,i}) \qquad (2.25)$$

In contrast, the heat transfer to the air in the cylinder is different with respect to each discrete event mentioned above. First, as the intake valve is open, that is, $\theta_i \in [0, \theta_{c,i})$, the heat transfer from the intake port to the cylinder with the intake air is formulated as follows:

$$\dot{q}_{c,i} = c_p T_{r,i} \dot{m}_{c,i} \qquad (2.26)$$

by considering that the intake port temperature $T_{r,i}$ is constant. Furthermore, assume that the cylinder is adiabatic from compression starting to the spark timing (i.e., $\theta_i \in [\theta_{c,i}, \theta_{s,i})$), as well as from combustion completing to the time of the outtake valve opening (i.e., $\theta_i \in [\theta_{b,i}, \theta_{e,i})$). Hence, there is no heat transfer during these periods. In other words,

$$\dot{q}_{c,i} = 0, \quad \forall \theta_i(t) \in [\theta_{c,i}, \theta_{s,i}), \text{ and } \theta_i \in [\theta_{b,i}, \theta_{e,i}) \qquad (2.27)$$

Same as for the intake stroke, during the exhaust stroke $\theta_i \in [\theta_{e,i}, 720)$, the heat transfer from the cylinder to the exhaust pipe with the exhaust can be formulated as

$$\dot{q}_{c,i} = -c_p T_{c,i} \dot{m}_{e,i} \qquad (2.28)$$

where $\dot{m}_{e,i}$ denotes the air mass flow rate passing through the exhaust valve of cylinder i. Moreover, the modeling for $\dot{m}_{e,i}$ is the same as for getting the model (2.15) by using the in-cylinder pressure and the exhaust manifold pressure.

As described above, except the combustion process, there is heat transfer to the air in the cylinder, and also, there is heat taken away with the exhaust. Besides these, there is no heat transfer. Modeling these phenomena is not difficult. However, during the combustion process, that is, $\theta_i \in [\theta_{s,i}, \theta_{b,i})$, it is not easy to deal with the heat transfer. In general, the heat-transfer rate, that is, the heat-release rate during combustion, is modeled using an experimental approach. Of course, this heat release determined by the air–fuel ratio (A/F), compression state, combustion efficiency, and so forth, can also be approximated with the Wiebe function [8]. For the sake of simplicity, another approximated function obtained by experimental data fitting is given as follows [9]:

$$dq_{c,i} = C_i(\theta_i, \theta_{s,i}, \theta_{b,i}, Q_{c,i}, a_i)d\theta \qquad (2.29)$$

where the function $C_i(\cdot)$ is described by

$$C_i(\cdot) = \frac{a_i(a_i+1)}{\theta_{s,i}^{a_i+1}} Q_{c,i}(\lambda_i)(\theta_i - \theta_{s,i})^{a_i-1}(\theta_{b,i} - \theta_i), \quad \theta_i \in [\theta_{s,i}, \theta_{b,i}) \qquad (2.30)$$

and the heat $Q_{c,i}$ that depends on the A/F λ_i in cylinder i is given by

$$Q_{c,i} = 41{,}868(-5.137\lambda_i^2 + 145.31\lambda_i - 421.69)(m_{c,i} + m_{f,i})$$

with the fuel mass $m_{f,i}$ and the air charge mass $m_{c,i}$ calculated by

$$m_{c,i} = \int_{t_{\theta_i=0}}^{t_{\theta_i=\theta_{c,i}}} \dot{m}_{c,i} dt$$

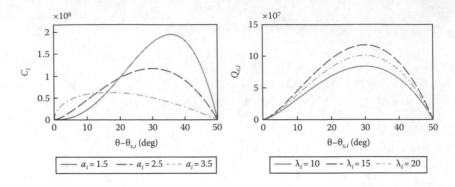

FIGURE 2.6
The heat-release rate.

The involved A/F λ_i in cylinder i is usually defined as

$$\lambda_i = \frac{\dot{m}_{c,i}}{\dot{m}_{f,i}} \tag{2.31}$$

while another definition used in the related engine research is given as

$$\lambda_i = \frac{m_{c,i}}{m_{f,i}} \tag{2.32}$$

Figure 2.6 shows the curves of $C_i(\cdot)$ and $Q_{c,i}$ when $m_{c,i} + m_{f,i}$ is constant. As shown in this figure, the coefficient $a_i(>0)$ actually plays the role to express the combustion speed.

Based on the above arguments and the energy conservation law (2.2), the in-cylinder pressure dynamics can be deduced as follows by sequently considering each discrete event:

$$\dot{p}_{c,i} = -\frac{\kappa}{V_{c,i}} p_{c,i} \dot{V}_{c,i} + \frac{\kappa}{V_{c,i}} \delta_{p,i}(T_{r,i}, T_{c,i}, \dot{m}_{c,i}, \dot{m}_{e,i}) \tag{2.33}$$

where the function $\delta_{p,i}(T_{r,i}, T_{c,i}, \dot{m}_{c,i}, \dot{m}_{e,i})$ is a switching function depending on the crank angle θ_i given as

$$\delta_{p,i} = \begin{cases} RT_{r,i}\dot{m}_{c,i}, & 0° \leq \theta_i < \theta_{c,i} \\ 0, & \theta_{c,i} \leq \theta_i < \theta_{s,i} \\ \frac{\kappa - 1}{\kappa} C_i(\theta_i, \theta_{s,i}, \theta_{b,i}, Q_{c,i}, a_i)\dot{\theta}_i, & \theta_{s,i} \leq \theta_i < \theta_{b,i} \\ 0, & \theta_{b,i} \leq \theta_i < \theta_{e,i} \\ -RT_{c,i}\dot{m}_{e,i}, & \theta_{e,i} \leq \theta_i < 720° \end{cases}$$

Figure 2.7 shows the profiles of in-cylinder pressure obtained from the real engine system data.

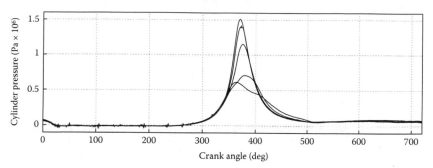

FIGURE 2.7
Variations in cylinder pressure with respect to different spark timings.

Furthermore, notice that the variation of internal energy in the cylinder depending on the temperature and the mass can be described as

$$du_i = c_v d(m_{c,i} T_{c,i})$$

Then, for each interval mentioned above, applying the first law of thermodynamics yields the in-cylinder temperature dynamics as follows:

$$\dot{T}_{c,i} = -\frac{p_{c,i}}{c_v m_{c,i}} \dot{V}_{c,i} + \frac{1}{c_v m_{c,i}} \delta_{T,i}(T_{r,i}, T_{c,i}, \dot{m}_{c,i}, \dot{m}_{e,i}) \qquad (2.34)$$

Here, similar to the pressure dynamics (2.33), the switching function $\delta_{T,i}(T_{r,i}, T_{c,i}, \dot{m}_{c,i}, \dot{m}_{e,i})$ is given as

$$\delta_{T,i}(\cdot) = \begin{cases} c_p T_{r,i} \dot{m}_{c,i} - c_v T_{c,i} \dot{m}_{c,i}, & 0° \le \theta_i < \theta_{c,i} \\ 0, & \theta_{c,i} \le \theta_i < \theta_{s,i} \\ C_i(\theta_i, \theta_{s,i}, \theta_{b,i}, Q_{c,i}, a_i)\dot{\theta}_i, & \theta_{s,i} \le \theta_i < \theta_{b,i} \\ 0, & \theta_{b,i} \le \theta_i < \theta_{e,i} \\ -RT_{c,i}\dot{m}_{e,i}, & \theta_{e,i} \le \theta_i < 720° \end{cases}$$

In this function, the change rate of the in-cylinder air mass $m_{c_i}(t)$ is characterized by (2.19) during the intake stroke, and for other strokes, it can be expressed by

$$\dot{m}_{c,i} = \begin{cases} 0, & \theta_{c,i} \le \theta_i < \theta_{e,i} \\ -\dot{m}_{e,i}, & \theta_{e,i} \le \theta_i < 720° \end{cases} \qquad (2.35)$$

The in-cylinder air volume $V_{c,i}$ is changed following the rotational motion of the engine. Using the parameter definitions illustrated in the crankshaft system, this volume can be given as follows.

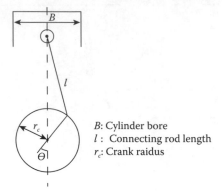

B: Cylinder bore
l : Connecting rod length
r_c: Crank raidus

FIGURE 2.8
Parameter definitions in the crankshaft system.

Figure 2.8 illustrates the parameter definitions in the crankshaft system. First, the piston displacement x and the crank angle are related by

$$
\begin{aligned}
x(\theta) &= r_c + l - \left(r_c \cos\theta + \sqrt{l^2 - r_c^2 \sin^2\theta} \right) \\
&= r_c(1 - \cos\theta) + l - \sqrt{l^2 - r_c^2 \sin^2\theta}
\end{aligned}
\tag{2.36}
$$

Denote $V_{l,i}$ as the minimum cylinder volume at TDC. With the displacement (2.36), the in-cylinder air volume of cylinder i can be given by

$$
\begin{aligned}
V_{c,i}(\theta) &= \frac{\pi B^2}{4} x_i(\theta) + V_{l,i} \\
&= \frac{\pi B^2}{4} r_c \left\{ 1 - \cos[\theta - \pi(i-1)] \right\} + l - \sqrt{l^2 - r_c^2 \sin^2[\theta - \pi(i-1)]} + V_{l,i}
\end{aligned}
\tag{2.37}
$$

Finally, let $A_{c,i}$ denote the effective area of the piston. The force in cylinder i acting to move the piston can be expressed as

$$
F_{c,i} = (p_{c,i} - p_a) A_{c,i}
\tag{2.38}
$$

Rotational Motion of Crankshaft. The generated in-cylinder pressure gives the force exerted on the piston, which produces the torque for the rotational motion of the crankshaft through the connection rod. Dealing with the piston, connecting rod, and crankshaft as a lumped mass system, the mechanical movement equation is deduced as below. See Figure 2.9 for an illustration of the involved parameters and forces of the system.

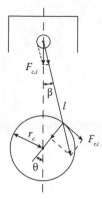

FIGURE 2.9
The force on the crankshaft.

Calculating the derivative of the piston displacement (2.36) along the crank angle yields

$$\frac{dx(\theta)}{d\theta} = r_c \left(\sin\theta + \frac{\cos\theta\sin\theta}{\sqrt{\left(\frac{l}{r_c}\right)^2 - \sin^2\theta}} \right) \tag{2.39}$$

Suppose that $(l/r_c)^2 \gg \sin^2\theta$; it can be approximated to

$$\frac{dx(\theta)}{d\theta} = r_c \left(\sin\theta + \frac{r_c\sin 2\theta}{2l} \right) \tag{2.40}$$

Then, the moving speed of the piston can be obtained as

$$v(t) = \frac{dx(\theta)}{d\theta}\dot\theta = \dot\theta r_c \left(\sin\theta + \frac{r_c\sin 2\theta}{2l} \right) \tag{2.41}$$

from which the inertial force can be obtained below:

$$F_m = M\frac{dv}{dt} = M\frac{dv}{d\theta}\dot\theta = Mr_c\dot\theta^2 \left(\cos\theta + \frac{r_c\cos 2\theta}{l} \right) \tag{2.42}$$

where M denotes the mass of the crankshaft. Combining with the gravity F_g, the resultant force in the tangent direction of the crank rotation is given by the following equation:

$$F_{r,i} = (F_{c,i} + F_m + F_g)\cos\beta\sin(\theta + \beta) \tag{2.43}$$

Then, the torque acting on the crankshaft is

$$\tau_{e,i} = r_c F_{r,i} \tag{2.44}$$

and using the relationship $l \sin \beta = r_c \sin \theta$, the torque can be rewritten as

$$
\tau_{e,i} = (F_{c,i} + F_m + F_g)r_c \left\{ \sin \theta \left(1 - \frac{r_c^2 \sin^2 \theta}{l^2} \right) + \frac{r_c}{l} \cos \theta \sin \theta \sqrt{1 - \frac{r_c^2 \sin^2 \theta}{l^2}} \right\}
$$

$$(2.45)$$

Furthermore, taking into account the load torque τ_l and the friction torque τ_f, the rotational movement equation of the crankshaft can be represented by

$$
J\ddot{\theta} = \sum_{i=1}^{n} \tau_{e,i} - \tau_l - \tau_f
$$

$$(2.46)$$

where J denotes the inertial moment of the crankshaft.

2.5   Mean-Value Model

The complexity of the engine model introduced in Section 2.4 is due to considerably focusing on the characteristics of each cylinder and each intake pipe. Compared to the volume of the intake manifold, the volume of the intake pipe is negligibly small, and modeling of each cylinder results in nonsmooth representation of the air mass flow rate $\dot{m}_o$. This causes the nonsmooth representation of the engine torque acting on the crankshaft. Neglecting this specific characteristic of the intake pipe, the simplified so-called mean-value engine model is obtained by averaging the air mass flow and engine torque equivalently.

First, consider the air mass flow entering each cylinder from the manifold $\dot{m}_o$. Let m_o denote the air charge of each cylinder in one cycle, and assume that it can be accumulated in the manifold with volume V_o. Furthermore, suppose that its pressure and temperature are the same as those of the states of the manifold. Then applying the ideal gas law,

$$
p_m V_o = R m_o T_m
$$

$$(2.47)$$

gives the following relationship:

$$
m_o = \frac{V_o p_m}{R T_m} = \frac{\eta V_c p_m}{R T_m}
$$

$$(2.48)$$

where V_c denotes the displaced volume of the engine and $\eta = V_o/V_c$ is called the volumetric efficiency.

Meanwhile, for the air with mass m_o, if its volume is V_a at the manifold temperature T_m, by considering that its pressure and density are taken as the atmosphere pressure p_a and density ρ_a, one obtains the following equation:

$$V_a p_a = R m_o T_m = R \rho_a V_a T_m \tag{2.49}$$

which combining with Equation 2.48 yields

$$m_o = \frac{\rho_a \eta V_c}{p_a} p_m \tag{2.50}$$

Let ω (rad/s) denote the engine speed. Then the time of one cycle equals $4\pi/\omega$ (s). The cycle-based mean-value air mass flow rate can be calculated as

$$\dot{m}_o = \frac{\rho_a \eta V_c}{4\pi p_a} p_m \omega \tag{2.51}$$

With this mean-value airflow, the model of manifold pressure and temperature (Equations 2.12 and 2.13) can be rewritten as

$$\dot{p}_m = \frac{\kappa - 1}{V_m} \dot{Q}_m + \frac{\kappa R}{V_m} \left(T_a \dot{m}_{th} - T_m \frac{\rho_a \eta V_c}{4\pi P_a} p_m \omega \right) \tag{2.52}$$

$$\dot{T}_m = \frac{1}{c_v m} \dot{Q}_m + \frac{\kappa T_a - T_m}{m} \dot{m}_{th} + (1 - \kappa) \frac{T_m}{m} \cdot \frac{\rho_a \eta V_c}{4\pi P_a} p_m \omega \tag{2.53}$$

For further simplification of the model, suppose that the manifold is isothermal, that is, $\dot{Q}_m = \dot{T}_m = 0$. Then, Equation 2.13 gives the following relationship:

$$\kappa(T_a \dot{m}_{th} - T_m \dot{m}_o) = T_m(\dot{m}_{th} - \dot{m}_o)$$

Under the above assumption, the manifold dynamics can be represented by the following one-state model:

$$\dot{p}_m = \frac{R T_m}{V_m} \dot{m}_{th} - \frac{R T_m}{V_m} \cdot \frac{\rho_a \eta V_c}{4\pi p_a} p_m \omega \tag{2.54}$$

Moreover, as introduced in Section 2.4, the force that acts on the piston is calculated from the individual cylinder pressure and provides the engine torque. As illustrated by Equations 2.38, 2.45, and 2.46, engine torque is represented by a nonsmooth function that depends on the crank angle and cylinder pressure. Based on the cycle-based mean-value air mass flow rate, the engine torque can also be represented in the mean-value treatment.

Indeed, denoting Q as the heat release from a unit mass of air with complete combustion at standard conditions, the heat release from the combustion of the air mass dm_o is $Q dm_o$. The engine system is an equipment that transfers this thermal energy into mechanical energy, which generates the rotation $d\theta$ to the crankshaft. The actual mechanical energy of the engine system can be represented as $\tau_e d\theta$. Moreover, as introduced in Section 2.2, the engine torque generation starts after the compression stroke that follows the intake stroke. Considering this time delay, the so-called intake-to-power stroke delay, denoted as t_d, and denoting the conversion efficiency from thermal energy to mechanical energy as c_f, the following energy balance can be obtained:

$$\tau_e(t)d\theta(t) = c_f Q dm_o(t - t_d) \tag{2.55}$$

which means that

$$\tau_e(t) = c_f Q \frac{dm_o(t - t_d)}{d\theta(t)} = c_f Q \frac{\dot{m}_o(t - t_d)}{\omega} \tag{2.56}$$

Combining with the mean-value model of air mass flow rate (2.51) and noting $\omega(t - t_d) \simeq \omega(t)$, the mean-value torque generation can be described by the following equation:

$$\tau_e = \frac{\rho_a c_f \eta Q V_c}{4\pi p_a} p_m(t - t_d) \tag{2.57}$$

Regarding the efficiency c_f, it may change depending on the conditions, such as the A/F and spark timing at the current cycle.

The friction torque is modeled as [10]

$$\tau_f = D\omega + D_0 \tag{2.58}$$

where D and D_0 are constants.

Finally, rearranging the crankshaft rotational dynamics (2.46) by considering the above engine torque and friction torque, the following dynamic model deduced in a cycle-based mean-value sense can characterize the rotational movement of the engine:

$$\begin{cases} \dot{\omega}(t) = a_1 p_m(t - t_d) - \bar{D}\omega(t) - T_D \\ \dot{p}_m(t) = a_u u_{th}(t) - a_2 p_m(t)\omega(t) \end{cases} \tag{2.59}$$

with the parameters defined as follows:

$$a_1 = \frac{c_f \rho_a V_c \eta Q}{4\pi J p_a}, \quad \bar{D} = \frac{D}{J}, \quad T_D = \frac{D_0 + \tau_l}{J}$$

$$a_u = \frac{R T_m}{V_m} A_{th}\left(\frac{\pi}{2}\right) \psi(p_m, p_a, T_a), \quad a_2 = \frac{R T_m}{V_m} \cdot \frac{\rho_a V_c \eta}{4\pi p_a}$$

where the control input u_{th}, which can be realized by tuning the throttle opening ϕ, is defined as

$$u_{th}(t) := 1 - \cos\phi$$

and $\psi(p_m, p_a, T_a)$ is the function given by (2.17).

2.6 Numerical Simulation Model

As an example, Figure 2.10 shows an overall structure of the model for a four-cylinder engine. To illustrate the internal signal flows of the engine dynamics, most state variables of each block are shown to indicate the relationship as clear as possible. Detailed operation blocks and static function relations have been omitted.

Figure 2.11 shows a numerical simulation result obtained by using the deduced engine model. The parameters in the simulator are not in accordance with a specific engine. The physical parameters of the engine system are listed in Table 2.1. This simulation is under the operation by giving step commands to the throttle opening. Figure 2.11a shows the response of the manifold pressure. Figure 2.11b shows the P-V diagram. As can be seen in Figure 2.11b, the P-V diagram shows two states with respect to two constants of the throttle

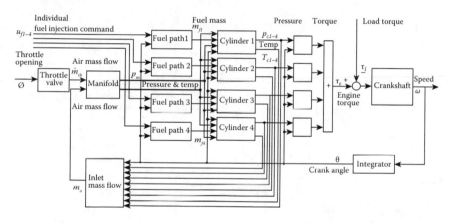

FIGURE 2.10
Structure of engine model.

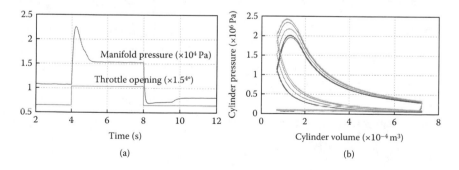

FIGURE 2.11
Simulation result. (a) Response of manifold pressure; (b) P-V diagram.

TABLE 2.1

Basic engine specifications

Number of cylinders	4
Bore (mm)	88.5
Stroke (mm)	95.8
Compression ratio	9.6
Crank radius (mm)	47.9
Connection rod length (mm)	150
Combustion chamber volume (mm^3)	68.5244
Single-cylinder volume (mm^3)	589.31
Active throttle diameter (mm)	47.6
Manifold volume (mm^3)	12,300
Intake pipe volume (mm^3)	1,179
Crankshaft inertia (kg $\cdot$ m^2)	0.11

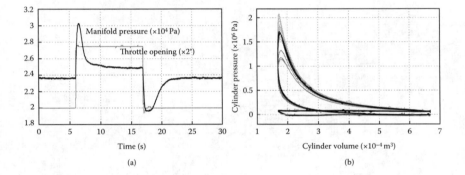

FIGURE 2.12

Experimental result obtained by a real engine test bench. (a) Response of manifold pressure; (b) *P-V* diagram.

opening. Moreover, compared to detailed numerical data, it is hopeful that this simulation result helps one to more easily understand the change of the cylinder pressure.

Figure 2.12 shows a corresponding experimental result conducted on an engine test bench. It can be observed by comparing the responses in Figures 2.11 and 2.12 that the proposed model captures the essential response characteristics of the real engine system.

2.7   Model Calibration Method

As discussed in Section 2.6, the parameters of the mean-value model are determined by the physics of the engine system. However, determining the values of the model parameter based on the physics does not guarantee the model accuracy because of the completed individual behavior of the cylinder and the mean-value treatment. Therefore, the model parameter is calibrated based on experiments for parameter identification.

Usually, to perform the model identification, the model of engine dynamics (2.59) is discretized with sampling rate T. Suppose that the engine operates around an equilibrium (ω_0, p_{m0}) and rewrite the crank rotational dynamics as

$$\dot{\tilde{\omega}} = a_1 \tilde{p}_m(t - t_d) - \bar{D}\tilde{\omega}$$

where $\tilde{\omega} = \omega - \omega_0$ and $\tilde{p}_m = p_m - p_{m0}$. In this case, $T_D = a_1 p_{m0} - \bar{D}\omega_0$. Then, the discrete-time engine model to be identified is as follows:

$$p_m(t_{k+1}) - p_m(t_k) = a_u T u_{th}(t_k) - a_2 T p_m(t_k)\omega(t_k) \tag{2.60}$$
$$\tilde{\omega}(t_{k+1}) = (1 - \bar{D}T)\tilde{\omega}(t_k) + a_1 T \tilde{p}_m(t_k) \tag{2.61}$$

where $t_k = kT$ $(k = 1, \cdots, n)$ denotes the sampling instant.

Under a command value as shown in Figure 2.13, engine speed has a dynamic regulation around 1,500 rpm. Collect the response data, including the throttle opening, manifold pressure, and engine speed, by the data acquisition system of the test bench. With these off-line data, a recursive least-squares (RLS) algorithm is applied to the dynamical models (2.60) and (2.61), respectively. Table 2.2 lists the identified parameters under two operation conditions of the engine with respect to the throttle opening. To evaluate the efficiency of the mean-value engine model, Figure 2.13 shows a validation result. In this experiment, the external load τ_l is set as 0, and the throttle opening command is a rectangular signal from 3.4° to 3.6°.

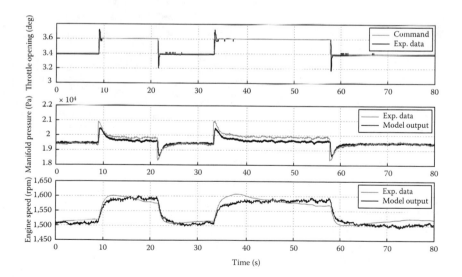

FIGURE 2.13
Experimental data versus model output.

TABLE 2.2

Identified parameters of mean-value engine model

Speed	Model Parameter				
ω/rpm	$a_1 \times 10^{-3}$	$\bar{D}$	T_D	$a_u \times 10^9$	a_2
1,500	7.8	0.208	118.97	3.38	1.94
2,000	6.8	0.107	128.237	3.88	2.61

From the identification results, it can be noted that engine models (2.59) with constant parameters are basically effective to characterize the dynamics under certain operating conditions with respect to engine speed. In fact, it is clear that physical parameters involved in the definition of the model parameters are dependent on engine variables under different operating modes, such as the volumetric efficiency that depends on the engine states (ω, p_m) extremely [14].

2.8 A Lemma: Air Mass Flow Passing through a Nozzle

The air mass flows that pass through the throttle and the intake and exhaust valves can be approximated by a stream of fluid passing through the cross section of a nozzle, as shown in Figure 2.14. Due to the changes of the throttle opening and the lifting of the intake and exhaust valves, the area for the air mass flows is changed. This situation can be considered to be a steady flow passing through a nozzle with an adjustable cross section. Then, the qualities of the air mass flow are the same at the cross sections.

Consider the volume encompassed by the dashed line shown in Figure 2.14a, where the surface area gradually converges. This part is called the control volume, and the right-hand side denotes the minimum cross section of the nozzle. In what follows, assume that the airflow in the control volume is one-dimensional, frictionless, adiabatic, and free of inertia influence. The control volume is illustrated by the image shown in Figure 2.14b, where the upstream conditions are denoted as area A_u, pressure p_u, velocity v_u, temperature T_u, and enthalpy h_u, and the downstream conditions are with area A_d, pressure p_d, velocity v_d, temperature T_d, and enthalpy h_d. By using the mass and energy conservation laws, the characteristics of the airflow in the control volume are shown as follows [7].

In order to describe the airflow characteristics, the following thermodynamic equations applying to an ideal gas are introduced first:

$$h_u = c_p T_u \qquad (2.62)$$

$$h_d = c_p T_d \qquad (2.63)$$

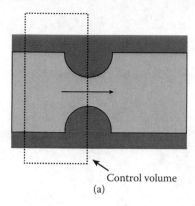
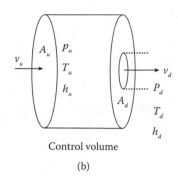

Control volume
(a)

Control volume
(b)

FIGURE 2.14
Air passing through a nozzle. (a) The image. (b) The control volume.

and under an adiabatic process, the following relationship of the ratio between the temperature and the pressure can be obtained. First, the absorbed heat dq of the air in the control volume is zero, that is, $dq = 0$. Then regarding the control volume, it can be obtained from the energy conservation law (2.2) that

$$du = -dw = -pdV \qquad (2.64)$$

Let T be the air temperature in the control volume and V be the volume of the control volume. Suppose that the air in the control volume is ideal gas. For unit air mass in the control volume, the following relationship can be obtained:

$$pV = RT \qquad (2.65)$$

Furthermore, considering that for ideal gas, $du = c_v dT$, submitting the ideal gas law (2.65) into Equation 2.64 gives

$$c_v dT = -pdV = -\frac{RT}{V} dV$$

which means that

$$\frac{dT}{T} = -\frac{R}{c_v} \frac{dV}{V}$$

Calculating the integral on the sides of the above equation, we get

$$\ln \frac{T_u}{T_d} = -\frac{R}{c_v} \ln \frac{V_u}{V_d}$$

and substituting the relationship $c_v = R/(\kappa - 1)$ given in (2.7) into the above equation obtains

$$\frac{T_d}{T_u} = \left(\frac{V_u}{V_d}\right)^{\kappa - 1}$$

With the above equation, the following relationship can be obtained by the ideal gas law (2.65):

$$\frac{T_d}{T_u} = \left(\frac{p_d}{p_u}\right)^{\frac{\kappa-1}{\kappa}} \tag{2.66}$$

Let m be the air mass in the control volume. For air mass flow $\dot{m}$, the mass conservation law gives

$$A_u \rho_u v_u = A_d \rho_d v_d = \dot{m} \tag{2.67}$$

and the energy conservation law gives

$$\frac{1}{2}v_u^2 + h_u = \frac{1}{2}v_d^2 + h_d \tag{2.68}$$

From the above relation, we have

$$v_d = \sqrt{v_u^2 + 2(h_u - h_d)} \tag{2.69}$$

Consider the upstream area $A_u \to \infty$; that is, the upstream velocity is negligible in comparison with the downstream velocity, which means $v_u = 0$. This leads to

$$v_d = \sqrt{2(h_u - h_d)} \tag{2.70}$$

Then, substituting the Equations 2.62 through 2.66, we obtain

$$v_d = \sqrt{RT_u}\sqrt{\frac{2\kappa}{\kappa-1}\left[1 - \left(\frac{p_d}{p_u}\right)^{\frac{\kappa-1}{\kappa}}\right]} \tag{2.71}$$

From the ideal gas law $p_d V = m R T_d$, where V denotes the volume of the control volume, we have

$$\rho_d = \frac{m}{V} = \frac{p_d}{RT_d} \tag{2.72}$$

Then, substituting (2.70), (2.72), and (2.66) into (2.9), we obtain

$$\dot{m} = A_d \rho_d v_d$$

$$= A_d \frac{p_d}{RT_d}\sqrt{RT_u}\sqrt{\frac{2\kappa}{\kappa-1}\left[1 - \left(\frac{p_d}{p_u}\right)^{\frac{\kappa-1}{\kappa}}\right]}$$

$$= A_d \frac{p_u}{\sqrt{RT_u}}\left(\frac{p_d}{p_u}\right)^{\frac{1}{\kappa}}\sqrt{\frac{2\kappa}{\kappa-1}\left[1 - \left(\frac{p_d}{p_u}\right)^{\frac{\kappa-1}{\kappa}}\right]} \tag{2.73}$$

Furthermore, for the airflow passing through a nozzle, it is known that when the air pressure drop is large enough at the cross section with minimum area in the nozzle, the fluid will reach sonic velocity. And, the pressure ratio between the upstream and the downstream is called the critical pressure ratio when the sonic velocity is achieved. Let $(p_d/p_u)_c$ denote the critical pressure ratio. The value of $(p_d/p_u)_c$ is found as follows.

At the cross section with minimum area in the nozzle, the sonic velocity is

$$v_{sonic} = \sqrt{\kappa \frac{p_d}{\rho_d}} \tag{2.74}$$

Under the critical pressure condition, that is, $v_{sonic} = v_d$, Equations 2.70 and 2.74, with Equations 2.66 and 2.72, give

$$\left(\frac{p_d}{p_u}\right)_c = \left(\frac{2}{\kappa+1}\right)^{\frac{\kappa}{\kappa-1}} \tag{2.75}$$

When the fluid achieves sonic velocity at the outlet of the control volume, the drops of the air pressure will not affect the air mass flow rate. In this case, with $p_d/p_u = (p_d/p_u)_c$, the equation (2.73) of the air mass flow rate is

$$\dot{m} = A_d \frac{p_u}{\sqrt{RT_u}} \sqrt{\kappa \left(\frac{2}{\kappa+1}\right)^{\frac{\kappa+1}{\kappa-1}}} \tag{2.76}$$

From the above discussion, the air mass flow rate passing through the nozzle can be formulated by

$$\dot{m} = A_d \frac{p_u}{\sqrt{RT_u}} \Psi\left(\frac{p_d}{p_u}\right) \tag{2.77}$$

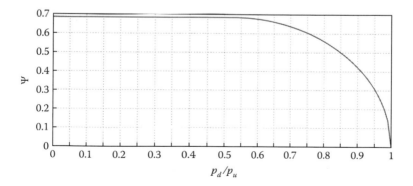

FIGURE 2.15
The influence function of the pressure ratio in a nozzle.

where

$$\Psi\left(\frac{p_d}{p_u}\right) = \begin{cases} \left(\dfrac{p_d}{p_u}\right)^{\frac{1}{\kappa}} \sqrt{\dfrac{2\kappa}{\kappa-1}\left[1-\left(\dfrac{p_d}{p_u}\right)^{\frac{\kappa-1}{\kappa}}\right]}, & \dfrac{p_d}{p_u} \geq \left(\dfrac{2}{\kappa+1}\right)^{\frac{\kappa}{\kappa-1}} \\[4mm] \sqrt{\kappa\left(\dfrac{2}{\kappa+1}\right)^{\frac{\kappa+1}{\kappa-1}}}, & \dfrac{p_d}{p_u} < \left(\dfrac{2}{\kappa+1}\right)^{\frac{\kappa}{\kappa-1}} \end{cases} \quad (2.78)$$

Consider the upstream of the nozzle is under the standard atmosphere pressure, that is, $p_u = 1.01 \times 10^5$ (Pa) and $\kappa = 1.4$. Then, the critical pressure ratio is 0.5283. Figure 2.15 shows the curve of the function (2.78) with respect to the pressure ratio.

3

Speed Control

3.1 Introduction

Engine speed control is a classical issue in automotive control applications. The performance of engine speed has significant impacts on vehicle design attributes such as comfort, emission, and fuel economy, especially during transitional operations [15, 16]. In the community of control engineering, this has led to many approaches to tackle the speed control problem, such as l_1 optimal control [17], sliding mode control [18], fuzzy control [15], adaptive control [19], model predictive control [20], and others referring to the references therein.

One of the main characteristics is the intake-to-power stroke delay t_d involved in modeling the engine dynamics in the continuous time domain introduced in Chapter 2. The delay time t_d is a time-varying variable dependent on the engine speed, and taking engines with six cylinders as an example, it can be calculated by

$$t_d = \frac{\pi}{\omega} \tag{3.1}$$

As is well known, the presence of time delay in the system may induce undesired behaviors of oscillation and instability; therefore, this delay characteristic should be considered properly in investigating engine control problems [21]. Several speed control methods that take the intake-to-power stroke delay into account have been proposed [22–24], in which the controllers are constructed by applying the design techniques for linear systems to the linearized engine model. On the other hand, the engine system, as an active benchmark example, has been used to assess the control design methods for time-delay systems [25, 26].

In this chapter, a nonlinear engine speed control strategy will be introduced. From the viewpoint of easily practical implementation, the proposed control scheme has a simple structure that consists of the speed error feedback with nonlinear proportional gain only. Furthermore, the presented scheme focuses on dealing with transient speed control; hence, a model that generates a reference trajectory for any desired speed is used to construct the feedback controller to ensure the tracking performance. Moreover, a nominal model-based feedforward compensation is appended to achieve a rapid responsibility. The stability analysis will be exactly provided based on the engine model involving the intake-to-power stroke delay. It will be shown that the proposed control system is asymptotically stable at the desired speed if the nonlinear gain function satisfies the provided condition. Then, the case that utilizes a traditional proportional controller with constant feedback gain will be discussed. In this case, a sufficient condition for the constant gain will be given such that the control system is locally asymptotically stable, and an estimated domain of attraction that plays a significant role from the view of practical applications will be provided. In addition to considering a constant intake-to-power stroke delay in the stability analysis, the case for a time-varying delay will also be discussed. The analysis results are obtained with the Lyapunov–Krasovskii functional stability theorem for functional differential equations. Furthermore, based on the obtained theoretical contribution, control schemes for speed control during the starting operating mode are proposed. Finally, experimental validation results are carried out for each control scheme.

Generally, the key issues in the engine speed control design can be considered as the following two aspects [27].

Speed set-point regulation: In many situations, such as for dealing with the basic problem of idling speed control in automotives and for energy management in hybrid electric vehicles, the reference engine speed should be properly set for yielding high engine performance in terms of fuel economy, emission, and so forth.

Load disturbance rejection: The rejection of the load disturbances is an essential task for the speed control system design.

The key characteristics of the proposed speed control schemes are presented as follows:

- A model-based design framework for the nonlinear time-delay engine system is developed; the control scheme structure is simple by using a feedforward compensation combined with a proportional-like feedback control law such that the control strategies are easily implementable.

- The control scheme focuses on attending good transient performance of the engine speed with an introduced filter.

- The convergence of the closed-loop control system is analyzed using the Lyapunov–Krasovskii stability theory. (A brief review is shown in Appendix B.)

- The control law performance that can be robust to changes in the modeling and to disturbances from imprecise known external loads is demonstrated.

3.2 Preliminaries

To investigate the speed control issues, preliminary discussions on the structures of the control system are presented here.

Based on the mean-value engine model (2.59) introduced in Chapter 2, for a given desired engine speed, if the engine system is stable at this speed, the internal states of the engine speed control system will reach a corresponding equilibrium. At a desired engine speed ω_r, let (ω_r, p_m^*) denote the equilibrium. It can be deduced that the equilibrium of the engine speed control system satisfies the following equations:

$$\begin{cases} 0 = a_1 p_m^* - \bar{D}\omega_r - T_D(\tau_l) \\ 0 = a_u u_{th}^* - a_2 p_m^* \omega_r \end{cases} \tag{3.2}$$

where the u_{th}^* denotes the control input with respect to the required throttle opening to keep the system states at equilibrium. Moreover, since the parameter T_D in the model (2.59) is dependent on the external load, when the load changes, with respect to the desired speed ω_r, u_{th}^*, and p_m^* have the relationship shown in Figure 3.1a. Moreover, according to the above discussion on equilibrium, it should be clear that in the case where the load torques are known, feedforward compensation is an effective approach to reject the load disturbances to the engine speed control system. According to the static map data related to the parameters of the mean-value engine model (2.59), the feedforward compensation can be designed as

$$u_{th}^* = \frac{1}{a_u} \left[a_2 \omega_r \cdot \frac{1}{a_1} (\bar{D}\omega_r + T_D(\tau_l)) \right] \tag{3.3}$$

It can be seen from (3.3) that when the engine operates at a static speed value ω_r, the feedforward compensation u_{th}^* is physically dependent only on the external load τ_l, as shown in Figure 3.1b, where a map obtained from the engine test bench is given.

Although the feedforward compensation is usually supposed to be necessary when it is available, there are many issues that it cannot address, for example, the improvement of speed transient performance in terms of overshoot and settling time. In this case, feedback control must be employed.

Basically for the case of a fixed air–fuel ratio (A/F) and spark timing, the engine speed control system is treated to as a single-input single-output (SISO) system with the throttle opening as control input and the speed as output. As introduced in Chapter 2, spark timing and the A/F performance can also affect the engine torque generation. Let $f_s(u_s) \in [0, 1]$ and $f_\lambda(\lambda)$ $(\in [0, 1])$ denote the normalized influences of spark timing and A/F to the

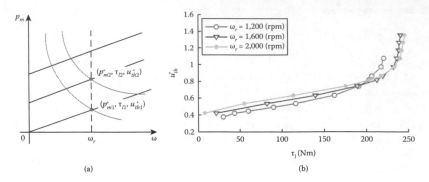

(a)   (b)

FIGURE 3.1
Operating points of the engine at steady states. (a) Relationship of the equilibrium p_m^* and u_{th}^* with respect to ω_r and τ_l. (b) Steady input values u_{th}^* under different load torques τ_l of a real engine.

engine torque generation, respectively. In this case, the mean-value engine torque generation (2.57) can be given as

$$\tau_e(t) = c_\tau p_m(t - t_d) f_s(u_s) f_\lambda(\lambda), \quad c_\tau = \frac{\rho_a c_f \eta Q V_c}{4\pi p_a} \tag{3.4}$$

In general, the A/F output of the engine system is a variable that is controlled to be the desired value separately by managing the control input of the fuel injection command. Hence, in the speed control loop, the influence from A/F to the torque generation is treated as an external disturbance. For the nominal case, the following condition is taken into consideration:

$$f_\lambda(\lambda) = 1 \tag{3.5}$$

For the influence from the spark timing, it actually characterizes the relation between the spark advance (SA) and the maximum spark advance for best torque (MBT). In the research, for example, [10] and [30], a nonlinear smooth function is proposed to model this influence:

$$f_s(u_s) = [\cos(u_s - M_{bt})]^{2.875} \tag{3.6}$$

where u_s denotes the control input of the SA command and M_{bt} denotes the MBT value. In addition to managing the throttle opening for speed control, spark timing control gives the other contribution. At a static engine operation (i.e., under constant speed and load), the SA command should be set at the MBT, which is experientially found. In other words, MBT should be defined as the feedforward control in the speed control loop:

$$u_s^* = M_{bt}(\omega_r, \tau_l) \tag{3.7}$$

However, it is well known that spark timing has a significant influence on the engine performance in terms of the emission. A trade-off should be consider in developing a speed control system.

Finally, combining (3.5) and (3.6) with (3.4), the mean-value engine torque generation can be rewritten as

$$\tau_e(t) = c_\tau p_m(t - t_d)[\cos(u_s - M_{bt})]^{2.875} \tag{3.8}$$

3.3 Idle Speed Control Scheme

The structure of a general engine speed control system can be summarized as in Figure 3.2. In the following sections, engine speed control schemes are proposed focusing on the designing of feedback controllers of the throttle opening. A case study considering two control inputs for speed control is provided in Section 3.5.

In this section, an idle speed control scheme will be introduced to deal with the speed set-point regulation. In [32], focusing on the normal engine system, a mean-value engine model-based design approach is presented. The case is that the SA is supposed to be at the corresponding MBT, and the A/F is at the desired value guaranteed by the fuel injection control loop of the engine system. The aim of the control strategy is to drive the engine speed to a given reference speed ω_r. Figure 3.3 shows the structure of the proposed control scheme.

According to the investigation on the calibration of the mean-value engine model in Section 2.7, the mean-value model (2.59) can characterize the engine dynamics with constant model parameters a_1, $\bar{D}$, a_u, and a_2 when the engine operates around a set idling speed. Moreover, assume that the

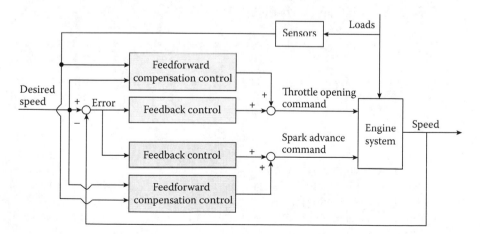

FIGURE 3.2
Structure of general engine speed control system.

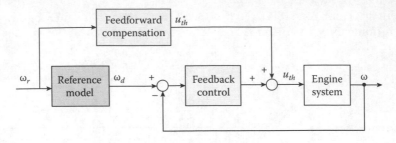

FIGURE 3.3
Structure of the speed control scheme II.

external load is measurable. In other words, the feedforward compensation
(3.3) is available for speed control design.

In order to achieve good transient speed performance, the speed transition
is forced to follow a desired speed trajectory ω_d generated by a first-order
filter, that is,

$$\dot{\omega}_d(t) = -\sigma(\omega_d(t) - \omega_r), \quad t \geq t_0 \tag{3.9}$$

with $\omega_d(t_0) = \omega(t_0)$, where parameter $\sigma > 0$ is a given constant and t_0 denotes
the initial timing of the speed tracking operation.

With the above reference model, the control input u_{th} is constructed by
combining the feedforward compensation (3.3) with a proportional feedback
controller:

$$u_{th} = u_{th}^* + k_p(\omega)e_\omega \tag{3.10}$$

where $k_p(\omega)$ is the feedback gain, which is dependent on the speed or can be
taken as a constant value, as deduced later, and

$$e_\omega = \omega_d - \omega$$

The control problems investigated below are to find the conditions of the
feedback gain function $k_p(\omega)$ and a constant k_p to achieve that the speed
tracking error

$$e_\omega(t) \to 0 \text{ as } t \to \infty, \quad \forall e_\omega(t_0) \tag{3.11}$$

And we consider that only engine speed is measurable for control design.

Based on the stability theory on time-delay systems introduced in
Appendix B, the following speed control results can be obtained.

Without loss of generality, the stability analysis of the control system
should be performed with the error dynamic system. Define

$$e_r = \omega_r - \omega_d, \quad e_p = p_m^* - p_m$$

Substituting the control law (3.10) with the relation (3.3) into the engine model (2.59) gives

$$\begin{cases} \dot{e}_r = -\sigma e_r \\ \dot{e}_\omega = (\sigma - \bar{D})e_r - \bar{D}e_\omega + a_1 e_p(t - t_d) \\ \dot{e}_p = -a_2 p_m^* e_r - (a_u k_p(\omega) + a_2 p_m^*)e_\omega - a_2 \omega e_p \end{cases} \quad (3.12)$$

Denote the state variables of the system as

$$x(t) = \begin{bmatrix} e_r(t) & e_\omega(t) & e_p(t) \end{bmatrix}^T$$

and $x_t = \begin{bmatrix} e_{rt} & e_{\omega t} & e_{pt} \end{bmatrix}^T$ denotes the corresponding state with delay time, that is, $x(t - t_d)$. If this error system is stable in the sense of Lyapunov stability, the speed tracking error should be bounded. Furthermore, if it is said to be asymptotical stable, the speed error should be demonstrated to converge to zero.

In fact, it is possible that using the following Lyapunov–Krasovskii functional can prove the above statement:

$$V(x_t) = \frac{\gamma_1}{2}e_r^2 + \frac{\gamma_2}{2}e_\omega^2 + \frac{1}{2}e_p^2 + \frac{1}{2}\int_{t-t_d}^{t} e_p^2(s)ds \quad (3.13)$$

where γ_1 and γ_2 are given by

$$\gamma_1 = \frac{1}{\sigma}\left(\frac{1}{2} + \frac{1}{2}a_2^2 p_m^* + \varsigma\right) \quad (3.14)$$

with any constant $\varsigma > 0$, and

$$\gamma_2 = \frac{\bar{D} + \sqrt{\bar{D}^2 - 2[(\sigma - \bar{D})^2 + a_1^2]\varepsilon}}{(\sigma - \bar{D})^2 + a_1^2} \quad (3.15)$$

with a given ε satisfying

$$0 < \varepsilon < \frac{\bar{D}^2}{2[(\sigma - \bar{D})^2 + a_1^2]} \quad (3.16)$$

which guarantees that $\gamma_1 > 0$ and $\gamma_2 > 0$.

In the analysis presented below, the condition

$$a_2 \omega > 1 \quad (3.17)$$

is employed, and in fact, it can be confirmed according to the model calibration results shown in Chapter 2 that this condition is true under any operating modes of the engine.

Case of a Nonlinear Feedback Gain Condition. For the sake of simplicity, the delay time t_d is treated as constant; that is, it takes the nominal value at the desired speed ω_r, $t_d = \pi/\omega_r$.

For the candidate $V(x_t)$ given by (3.13), the time derivative $\dot{V}(x_t)$ on the trajectory of system (3.12) can be calculated as

$$\dot{V}(x_t) = \frac{\partial V(x_t)}{\partial x_t(0)}\dot{x} + \frac{1}{2}(e_p^2 - e_{pt}^2) \tag{3.18}$$

Substituting (3.12) into the above equation gives

$$\dot{V}(x_t) = -\gamma_1 \sigma e_r^2 + \gamma_2(\sigma - \bar{D})e_r e_\omega - \gamma_2 \bar{D}e_\omega^2 + \gamma_2 a_1 e_\omega e_p(t - t_d) - a_2 p_m^* e_r e_p$$
$$- (a_u k_p + a_2 p_m^*)e_\omega e_p - a_2 \omega e_p + \frac{1}{2}e_p^2 - \frac{1}{2}e_p(t - t_d) \tag{3.19}$$

Then completing with a square to the above equation gives

$$\dot{V}(x_t) = -\gamma_1 \sigma e_r^2 + \frac{1}{2}e_r^2 - \left(\frac{1}{\sqrt{2}}e_r - \frac{\gamma_2(\sigma - \bar{D})}{\sqrt{2}}e_\omega\right)^2 + \frac{1}{2}\gamma_2^2(\sigma - \bar{D})^2 e_\omega^2$$
$$- \gamma_2 \bar{D}e_\omega^2 + \frac{1}{2}\gamma_2^2 a_1^2 e_\omega^2 - \left(\frac{1}{\sqrt{2}}\gamma_2 a_1 e_\omega - \frac{1}{\sqrt{2}}e_p(t - t_d)\right)^2 + \frac{1}{2}e_p^2(t - t_d)$$
$$+ \frac{1}{2}a_2^2 p_m^{*2} e_r^2 - \left(\frac{a_2 p_m^*}{\sqrt{2}}e_r + \frac{1}{\sqrt{2}}e_p\right)^2 + \frac{1}{2}e_p^2$$
$$- (a_u k_p + a_2 p_m^*)e_\omega e_p - a_2 \omega e_p + \frac{1}{2}e_p^2 - \frac{1}{2}e_p(t - t_d)$$
$$\leq -\left(\gamma_1 \sigma - \frac{1}{2} - \frac{1}{2}a_2^2 p_m^{*2}\right)e_r^2 - \left(\gamma_2 \bar{D} - \frac{1}{2}\gamma_2^2(\sigma - \bar{D})^2 - \frac{1}{2}\gamma_2^2 a_1^2\right)e_\omega^2$$
$$- (a_2 \omega - 1)e_p^2 - (a_u k_p + a_2 p_m^*)e_\omega e_p \tag{3.20}$$

In view of the conditions (3.14) and (3.15), we have

$$\dot{V}(x_t) \leq -\zeta e_r^2 - \varepsilon e_\omega^2 - (a_2 \omega - 1)e_p^2 - (a_u k_p + a_2 p_m^*)e_\omega e_p$$
$$= -x^T Q(\omega)x \tag{3.21}$$

where

$$Q(\omega) = \begin{bmatrix} \zeta & 0 & 0 \\ 0 & \varepsilon & -\frac{1}{2}(a_u k_p(\omega) + a_2 p_m^*) \\ 0 & -\frac{1}{2}(a_u k_p(\omega) + a_2 p_m^*) & a_2 \omega - 1 \end{bmatrix} \tag{3.22}$$

Now it is clear how to choose $k_p(\omega)$ such that the matrix $Q(\omega)$ is positive-definite for any ω. Let $k_p(\omega)$ be given as

$$k_p(\omega) = \rho(t)\frac{2}{a_u}\sqrt{\varepsilon(a_2 \omega - 1)} - \frac{a_2}{a_u}p_m^* \text{ with a given function } |\rho(t)| < 1, \forall t \geq t_0 \tag{3.23}$$

which guarantees that a sufficient small constant $\epsilon > 0$ can be found such that

$$Q(\omega) \geq \epsilon I \quad \Rightarrow \quad \dot{V}(x_t) \leq -\epsilon\|x\|^2 \tag{3.24}$$

Note from (3.2) that

$$p_m^* = \frac{\bar{D}\omega_r + T_D(\tau_l)}{a_1} \tag{3.25}$$

Substituting it into the condition (3.23) we have

$$k_p(\omega) = \rho(t)\frac{2}{a_u}\sqrt{\varepsilon(a_2\omega - 1)} - \frac{a_2(\bar{D}\omega_r + T_D(\tau_l))}{a_u a_1} \tag{3.26}$$

Moreover, it is clear that for the Lyapunov–Krasovskii functional (3.13), there exist continuous nondecreasing functions $\mu_i(s)(> 0, s > 0)$ and $\mu_i(0) = 0$ $(i = 1, 2)$ such that

$$\mu_1(\|x\|) \le V(x_t) \le \mu_2(\|x_t\|_c) \tag{3.27}$$

Finally, from the above discussion, the following result can be obtained by the Lyapunov–Krasovskii stability theorem (Theorem B.1).

Proposition 3.1 *For any given $\sigma > 0$, if the feedback gain $k_p(\omega)$ is given by Equation 3.26 with a given function $|\rho(t)| < 1$, $\forall t \ge t_0$, then for any initial condition $x_0(t_d) \in C_r$, x_t asymptotically converges to zero as $t \to \infty$.*

Case of a Constant Feedback Gain Condition. Consider the case that constant feedback gain k_p is used. In the following, let $\xi > 0$ be a constant that satisfies

$$|e_r| \le \xi, |e_\omega| \le \xi, \quad \text{and} \quad u_2(\omega_r - 2\xi) > 1 \tag{3.28}$$

In this case, choose a feedback gain k_p in the controller (3.10) as

$$k_p \in \left(-\frac{2}{a_u}\sqrt{\varepsilon(a_2\omega_r - \xi)} - \frac{a_2(\bar{D}\omega_r + T_D(\tau_l))}{a_u a_1}, \right.$$
$$\left. \frac{2}{a_u}\sqrt{\varepsilon(a_2\omega_r - \xi)} - \frac{a_2(\bar{D}\omega_r + T_D(\tau_l))}{a_u a_1} \right) \tag{3.29}$$

Then it is easy to obtain that the time derivative of $V(x_t)$ given by (3.83) along the trajectory of system (3.12) satisfies

$$\dot{V}(x_t) \le -x^T Q' x$$

where

$$Q' = \begin{bmatrix} \zeta & 0 & 0 \\ 0 & \varepsilon & -\frac{1}{2}(a_u k_p + a_2 p_m^*) \\ 0 & -\frac{1}{2}(a_u k_p + a_2 p_m^*) & a_2(\omega_r - \xi) - 1 \end{bmatrix}$$

Taking into account the condition (3.29), we obtain that Q' is positive-definite, which implies that there exists a sufficiently small $\epsilon > 0$ such that

$$\dot{V}(x_t) \le -\epsilon x^T Q' x \tag{3.30}$$

This holds for any x_t in a domain $\mathcal{D}$ defined as

$$\mathcal{D} = \{x_t(t_d) \in C_r \mid |e_r| \leq \xi \ \& \ |e_\omega| \leq \xi\} \tag{3.31}$$

On the other hand, we can have that

$$V(x_t) \leq \mu_2(\|x_t\|_c) \tag{3.32}$$

$$\forall \ x_t \in \Omega = \left\{ x_t \in C_r \ \middle| \ \|e_r\|_c^2 + \frac{\gamma_2}{\gamma_1}\|e_\omega\|_c^2 + \frac{1+t_d}{\gamma_1}\|e_p\|_c^2 \leq \xi^2 \right\} \tag{3.33}$$

$$\text{with} \ \ \mu_2(\|x_t\|_c) = \frac{\|e_{rt}\|_c^2}{\xi^2} + \frac{\gamma_2\|e_{\omega t}\|_c^2}{\gamma_1\xi^2} + \frac{(1+t_d)\|e_{pt}\|_c^2}{\gamma_1\xi^2} \tag{3.34}$$

Then the following result can be obtained.

Proposition 3.2 *For any given ω_r and $\sigma > 0$, if the feedback gain k_p satisfies the condition (3.29), then the system (3.12) is local asymptotic stable at the origin over the domain $\mathcal{D}$ defined by (3.31). Furthermore, $\forall x_0(t_d) \in \Omega$, $x_t \to 0$ as $t \to \infty$ with the set Ω is given by (3.34).*

Proof. Combining with conditions (3.27) and (3.30), the local asymptotic stability of system (3.12) at the origin follows by the Lyapunov–Krasovskii stability theorem (B.1). On the other hand, it is clear that the set Ω defined by (3.34) is bounded and

$$\Omega \subset \mathcal{D} \tag{3.35}$$

Then, it follows from the condition (3.30) that

$$\dot{V}(x_t) \leq -\epsilon \left[\mu_2^{-1}(V(x_t)) \right]^2 \tag{3.36}$$

According to Theorem B.2, this implies that for any initial condition $x_0(t_d) \in \Omega$,

$$V(x_t) \to 0 \ \text{as} \ t \to \infty \ \Rightarrow \ x_t \to 0 \ \text{as} \ t \to \infty$$

This concludes the proof.

Condition for Case of Time-Varying Delay. In the propositions presented above, the delay time t_d is treated as constant, while it is determined by engine speed as π/ω exactly. If this time variability is taken into account, the time derivative of $V(x_t)$ will depend on dt_d/dt. However, as can be seen below, if t_d does not vary much quickly, then the stability of the error system can also be guaranteed by the same Lyapunov–Krasovskii functional with a slightly modified coefficient.

In fact, since in Proposition 3.1 ε satisfies condition (3.16), there exists a δ satisfying $0 < \delta < 1$ such that

$$\frac{2\varepsilon[(\sigma - \bar{D})^2 + a_1^2]}{1 - \bar{D}^2} = 1 - \delta \tag{3.37}$$

Let $M < \delta$. Then, define a constant γ_2' that satisfies $\gamma_2' > 0$ as

$$\gamma_2' = \frac{\bar{D} + \sqrt{\bar{D}^2 - 2\left[(\sigma - \bar{D})^2 + \dfrac{a_1^2}{1 - M}\right]\varepsilon}}{(\sigma - \bar{D})^2 + a_1^2} \tag{3.38}$$

We now choose a candidate of the Lyapunov–Krasovskii functional as

$$V'(x_t) = \frac{\gamma_1}{2}e_r^2 + \frac{\gamma_2'}{2}e_\omega^2 + \frac{1}{2}e_p^2 + \frac{1}{2}\int_{t-t_d}^t e_p^2(s)\,ds \tag{3.39}$$

where, along the trajectory of the error system (3.12) with feedback gain $k_p(\omega)$ (3.26) given by Proposition 3.1, we have

$$\dot{V}'(x_t) = -\gamma_1\sigma e_r^2 + \gamma_2'(\sigma - \bar{D})e_r e_\omega - \gamma_2'\bar{D}e_\omega^2 + \gamma_2'a_1 e_\omega e_p(t - t_d) - a_2 p_m^* e_r e_p$$
$$- (a_u k_p + a_2 p_m^*)e_\omega e_p - a_2 \omega e_p + \frac{1}{2}e_p^2 - \frac{1}{2}\left(1 - \frac{dt_d}{dt}\right)e_p(t - t_d) \tag{3.40}$$

Then with the following completed with a square,

$$\gamma_2'a_1 e_\omega e_p(t - t_d) - \frac{\gamma_2'^2 a_1^2}{2(1 - M)}e_\omega^2 - \left(\frac{\gamma_2'a_1}{\sqrt{2(1-M)}}e_\omega - \sqrt{\frac{1-M}{2}}e_p(l - l_d)\right)^2$$
$$+ \frac{1-M}{2}e_p^2(t - t_d)$$

the above time derivative (3.40) can be rearranged as

$$\dot{V}_1'(x_t) \leq -\left(\gamma_1\sigma - \frac{1}{2} - \frac{1}{2}a_2^2 p_m^{*2}\right)e_r^2 - \left(\gamma_2'\bar{D} - \frac{1}{2}\gamma_2'^2(\sigma - \bar{D})^2 - \frac{\gamma_2'^2 a_1^2}{2(1-M)}\right)e_\omega^2$$
$$- (a_2\omega - 1)e_p^2 - (a_u k_p + a_2 p_m^*)e_\omega e_p - \frac{1}{2}\left(M - \frac{dt_d}{dt}\right)e_p^2(t - t_d) \tag{3.41}$$

and due to (3.38), the above inequality becomes

$$\dot{V}'(x_t) \leq -x^T Q(\omega)x - \frac{1}{2}\left(M - \frac{dt_d}{dt}\right)e_p^2(l - t_d) \tag{3.42}$$

which means that if $dt_d/dt \leq M$, then

$$\exists\, \epsilon' > 0 \text{ such that } \dot{V}(x_t) \leq -\epsilon'\|x\| \tag{3.43}$$

The above discussion gives the following result guaranteed by the Lyapunov–Krasovskii stability theorem (B.1).

Corollary 3.1 *Consider systems (2.59) and (3.9) with controller (3.10) given by Proposition 3.1. There exists a δ satisfying $0 < \delta < 1$ such that the error system (3.12) is asymptotically stable at the origin if the delay time t_d satisfies*

$$\frac{dt_d}{dt} \leq M \tag{3.44}$$

where $M < \delta$.

3.4   Uncertainty and Robustness

In Section 3.3, two control results are presented to deal with the speed set-point regulation issue. The results are demonstrated to be effective under nominal engine operations. However, parameter perturbations of physical systems are generally inevitable. Regarding the engine speed control system, the parameter values of the control-oriented engine model can be varied and the accessory loads cannot be precisely known due to changes in engine operating modes and conditions. It can be noted that with these possible variations, the proposed feedforward compensation will be with error and the stability performance of the feedback controller may be lost. Hence, an implementable speed control law should be effective when the precise engine model is a mismatch and the load disturbance is not precisely known. In other words, a robust control law is essential in real implementations. In the following, two cases on robust control design will be introduced.

Extended Result of Proposition 3.1. For a given ϵ according to (3.16), a positive number q can be found such that

$$\sqrt{\epsilon(a_2\omega - 1)} \geq q \tag{3.45}$$

Moreover, there exists a positive constant Δ_m such that a constant M^0 can be found to guarantee that

$$a_u \left| \frac{a_2}{a_u} \cdot \frac{\bar{D}\omega_r + T_D(\tau_l)}{a_1} - M^0 \right| \leq \Delta_m \tag{3.46}$$

Then the following result can be obtained with the same Lyapunov–Krasovskii functional candidate (3.83).

Proposition 3.3 *Consider the control system (3.12). For any given reference speed ω_r and $\sigma > 0$, if the feedback gain $k_p(\omega)$ is designed as*

$$k_p(\omega) = \rho'(t)\frac{2}{a_u}\sqrt{\epsilon(a_2\omega - 1)} - M^0 \tag{3.47}$$

with the constant M^0 satisfying the condition (3.46), where Δ_m satisfies $\Delta_m < 2q$ and the continuous function $\rho'(t)$ satisfies

$$|\rho'(t)| < 1 - \frac{\Delta_m}{2q}, \quad \forall t \geq t_0 \tag{3.48}$$

then the origin is asymptotically stable and the engine speed ω converges to the reference speed ω_r from any initial condition $x_{t_0} \in C_r$.

Proof. For the feedback gain (3.47), the following condition can be obtained immediately in view of (3.46):

$$\left| k_p(\omega) + \frac{a_2(\bar{D}\omega_r + T_D(\tau_l))}{a_u a_1} \right|$$

$$= \left| \rho'(t)\frac{2}{a_u}\sqrt{\varepsilon(a_2\omega - 1)} - M^0 + \frac{a_2(\bar{D}\omega_r + T_D(\tau_l))}{a_u a_1} \right|$$

$$\leq |\rho'(t)|\frac{2}{a_u}\sqrt{\varepsilon(a_2\omega - 1)} + \Delta_m \tag{3.49}$$

Taking (3.48) and (3.45) into account, we obtain

$$\left| k_p(\omega) + \frac{a_2(\bar{D}\omega_r + T_D(\tau_l))}{a_u a_1} \right| < 2\left(1 - \frac{\Delta_m}{2q}\right)\sqrt{\varepsilon(a_2\omega - 1)} + \Delta_m$$

$$< 2\sqrt{\varepsilon(a_2\omega - 1)} \tag{3.50}$$

This guarantees that there exists a sufficiently small $\epsilon > 0$ such that the matrix $Q(\omega)$ in (3.22) with $k_p(\omega)$ given in (3.47) satisfies the condition (3.24). Therefore, the proof is achieved with the same argument used in Proposition 3.1.

A robust adaptive idle speed control scheme is shown below. Consider that the manifold pressure is measurable. The following analysis focuses on the unknown external load rejection under the situation that the model parameters are uncertain.

Motivated by the contribution in [29], a feedback compensation control law by SA is involved, and with the manifold pressure signal, this compensation is applied to deal with the intake-to-power delay. From the engine torque generation (3.8), the control law for SA can be designed as

$$u_s = M_{bt} + \arccos\left(\frac{p_m(t)}{p_m(t - t_d)}\right)^{\frac{1}{2.875}} \tag{3.51}$$

which guarantees that

$$\tau_e(t) = c_\tau p_m(t - t_d)[\cos(u_s - M_{bt})]^{2.875}$$

$$= c_\tau p_m(t) \tag{3.52}$$

With the above delay compensation (3.51), the engine system (2.59) for speed control design can be rewritten as

$$\begin{cases} \dot{\omega} = a_1 p_m - \bar{D}\omega - T_D(\tau_l) \\ \dot{p}_m = u_t - a_2 p_m \omega \end{cases} \tag{3.53}$$

where $u_t = a_u u_{th}$.

Then, for a given desired idle speed, the tracking dynamics between the desired trajectory and the trajectory of system (3.53) can be coordinated by the error $(e_\omega, e_p)^*$ as follows:

$$\begin{cases} \dot{e}_\omega = -\bar{D}e_\omega + a_1 e_p \\ \dot{e}_p = -u_t + a_2 \omega p_m \end{cases} \tag{3.54}$$

It should be noted that from (3.54), if all the model parameters are exactly known, then the speed control problem is trivial. In this case, by introducing the nonlinear compensation

$$u_t = a_2 p_m \omega + v \tag{3.55}$$

the poles of the second-order linear system can be assigned to any desired location by the new control signal v. However, the model parameters a_1, $\bar{D}$ and $T_D(\tau_l)$ are varied when the engine operating mode changes. This means that the error e_p is usually unmeasurable. Hence, the pole placement idea cannot be realized.

A design approach introduced in [33] shows that by taking the physics of the engine system into account, the desired speed controller with a simple structure can be designed. Considering that the model parameters a_1 and $\bar{D}$ are positive, the following analysis gives a desired feedback control law for the throttle opening with a simple structure. Design the feedback controller for throttle opening as

$$u_t = a_2 p_m \omega + k_p e_\omega \tag{3.56}$$

with a positive feedback gain k_p. Then, employing the candidate Lyapunov function,

$$V = \frac{k_p}{2} e_\omega^2 + \frac{a_1}{2} e_p^2 \tag{3.57}$$

we get that along the trajectory of the control system, its time derivative can be calculated as

$$\begin{aligned} \dot{V} &= k_p e_\omega(-\bar{D}e_\omega + a_1 e_p) + a_1 e_p(-k_p e_\omega) \\ &= -\bar{D} k_p e_\omega^2 \\ &\leq 0 \end{aligned} \tag{3.58}$$

This inequality (3.58) guarantees that the control system is stable at the equilibrium $(e_\omega = 0,\ e_p = 0)$ by the Lyapunov stability theory. Moreover, it

---

*External load rejection is focused; hence, the reference model is omitted in the control design, and the speed error is redefined to be $e_\omega := \omega_r - \omega$.

implies that the set $\Omega := \{(e_\omega, e_p) \mid \dot{V} = 0\} = \{e_\omega = 0\}$. Therefore, $e_\omega \to 0$ as $t \to \infty$ follows by the LaSalle invariance principle.

Note that by the above analysis, the control law (3.56) guarantees that for any unknown load disturbance τ_l, the desired engine speed is asymptotically stable. However, the control law (3.56) still relies on a condition that the exact model parameter a_2 should be known. Otherwise, the speed convergence cannot be achieved.

To extend the above control design to be model parameter independent, the following variables are introduced:

$$\hat{p}_m = p_m + \hat{e}_p \quad \text{and} \quad \tilde{e}_p = p_m^* - \hat{p}_m$$

With the above involved variable $\hat{p}_m$ and the related $\tilde{e}_p$, the speed control system can be represented by the following differential equations:

$$\begin{cases} \dot{e}_\omega = -\bar{D}e_\omega + a_1\tilde{e}_p + a_1\hat{e}_p \\ \dot{\hat{e}}_p = \dot{\hat{p}}_m - u_t + a_2\omega\hat{p}_m - a_2\omega\hat{e}_p \\ \dot{\tilde{e}}_p = -\dot{\hat{p}}_m \end{cases} \tag{3.59}$$

Then the following conclusion can be obtained.

Proposition 3.4 *Consider system (3.59). Let the feedback controller for the throttle opening be given by*

$$u_t = (1+\rho)k_p e_\omega + a_2\omega\hat{p}_m \tag{3.60}$$

with the following adaptive law:

$$\dot{\hat{p}}_m = \rho k_p e_\omega \tag{3.61}$$

where both k_p and ρ are any given positive constants. Then, the closed-loop system (3.59) with (3.60) is globally Lyapunov stable and for any constant load disturbance, the speed error $e_\omega \to 0$ as $t \to \infty$.

Proof. Choose the following candidate Lyapunov function to evaluate the control system:

$$V = \frac{k_p}{2a_1}e_\omega^2 + \frac{1}{2}\hat{e}_p^2 + \frac{1}{2\rho}\tilde{e}_p^2 \tag{3.62}$$

Along the trajectory of the system (3.59) with the control law (3.60), the time derivative of Lyapunov function (3.62) is

$$\dot{V} = \frac{k_p}{a_1}e_\omega(-\bar{D}e_\omega + a_1e_p + a_1\hat{e}_p) + \hat{e}_p(\dot{\hat{p}}_m - u_t + a_2\omega\hat{p}_m - a_2\omega\hat{e}_p) - \frac{1}{\rho}\tilde{e}_p\dot{\hat{p}}_m$$

$$= -k_p\frac{\bar{D}}{a_1}e_\omega^2 - a_2\omega\hat{e}_p^2 + \hat{e}_p(k_p e_\omega + \dot{\hat{p}}_m - u_t + a_2\omega\hat{p}_m) + \tilde{e}_p\left(k_p e_\omega - \frac{1}{\rho}\dot{\hat{p}}_m\right) \tag{3.63}$$

Taking the control designs (3.60) and (3.61) into account, the above derivative can be arranged as

$$\dot{V} = -k_p \frac{\bar{D}}{a_1} e_\omega^2 - a_2 \omega \hat{e}_p^2 \leq 0 \qquad (3.64)$$

This means that the control system by the controller (3.60) combining with (3.61) is stable by the Lyapunov stability theory. Moreover, since the set $\Omega := \left\{ (e_\omega, \hat{e}_p, \tilde{e}_p) \mid \dot{V} = 0 \right\} = \{ e_\omega = 0, \ \hat{e}_p = 0 \}$, the conclusion $e_\omega \to 0$ follows by the LaSalle invariance principle.

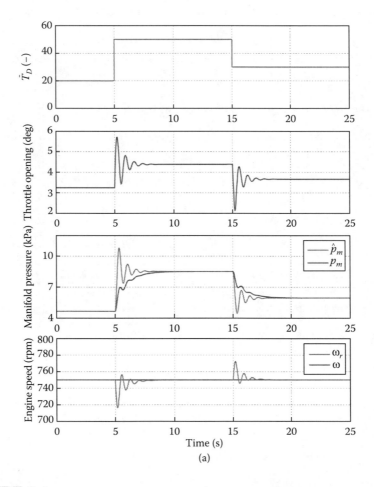

FIGURE 3.4
Simulation validation under controller (3.60) with adaptive law (3.61). (a) Result with known model parameter. *(Continued)*

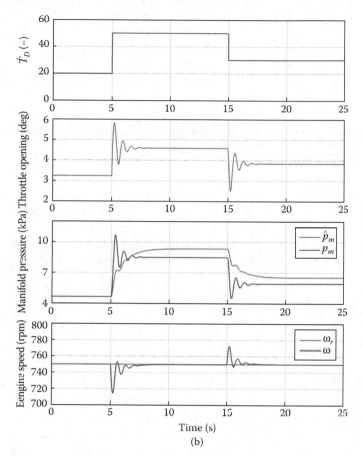

FIGURE 3.4 (Continued)
Simulation validation under controller (3.60) with adaptive law (3.61).
(b) Result with uncertain model parameter.

Remark 3.1 The above analysis actually demonstrates the speed convergence as well as $\hat{p}_m \to p_m$. And the controller (3.60) still needs the exact model parameter a_2. However, if an inaccuracy a_2 is used in the controller, when $e_\omega \to 0$, the controller satisfies $u_t \to \bar{a}_2 \hat{p}_m \omega$ (where $\bar{a}_2$ denotes a constant that is close to the precise value a_2), which can equivalently mean $u_t \to a_2 p_m \omega$. In this case, $e_\omega \to 0$, although $\hat{p}_m$ cannot converge to p_m. This observation is demonstrated by the following numerical simulation result shown in Figure 3.4. The result of Figure 3.4a is from testing under a normal system (i.e., the model parameters $\bar{D}$, a_1, a_u, and a_2 are constant during the simulation). The load torque-related parameter $\bar{T}_D(\tau_l)$ is given in two step changes. The result confirms the speed convergency under the load torque disturbance; moreover, it can be seen that $\hat{p}_m$ converges to p_m. In the testing for the result

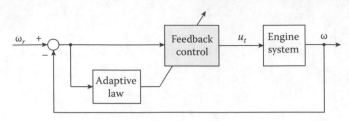

FIGURE 3.5
Structure of the robust adaptive speed control scheme.

of Figure 3.4b, the model parameter a_2 is given a 10% incensement at $t = 5$ s when $\bar{T}_D$ is changed. It can be found that the speed is asymptotic stable at the desired idle speed under load torque variation; however, $\hat{p}_m$ converges to a constant that is not the p_m, and it can also be verified that $u_t \to \bar{a}\hat{p}_m\omega$. This remark is also confirmed later by experimental testing.

Finally, the structure of the above robust adaptive speed control system is as shown in Figure 3.5.

3.5   Starting Speed Control

Starting is a typical transient operating mode of SI engines. The starting operation can generally breakup into four phases. Taking the V-type engine with six cylinders into consideration, Figure 3.6 shows a sketch of the starting control system. The crankshaft is initially driven to have a low constant rotational speed by the starting motor. Along the crank angle, the engine strokes of each cylinder follow. After the first ignition occurs in a cylinder, the driven torque on the crankshaft will be shifted from the starting motor to the engine. This will cause an acceleration of crank rotational velocity to achieve stable idle speed. The description shown above indicates that a basic requirement of engine starting is fast acceleration followed by quick convergence to a target idle speed.

To challenge the start control problem, a benchmark problem for the starting speed control was proposed by the Research Committee on Advanced Powertrain Control Theory of SICE in Japan. Details on the introduction can be found in Section 8.2. In this section, a challenging result is introduced.

The speed control system discussed in the previous sections is designed based on the control-oriented mean-value engine model. Moreover, the control loops for the commands of fuel injection and spark timing are considered to be idealized. However, during the starting, the engine characteristic changes rapidly. Following the transition of engine speed, manifold pressure will be down from the atmospheric pressure to a static value that is extremely far from its initial value. The walls of the intake manifold and cylinders will

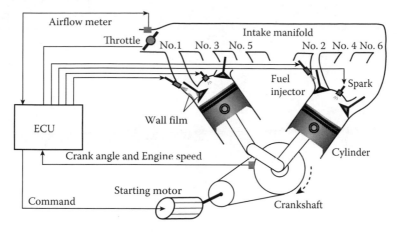

FIGURE 3.6
Sketch of engine starting control system.

be warmed up due to the combustion in the cylinders. All of these changes will affect the performance of engine starting and will certainly cause great limitations to use the mean-value engine models to describe the behaviors during starting. Especially in real engines, a combustion event in a cylinder can fail if the A/F is beyond a certain range (generally, the range is 12–18). In other words, engine behavior has the state transition between fire and misfire, depending on the A/F, as shown in Figure 8.8. As mentioned above, due to the significant changes of the characteristics in the fuel path and air intake path, fuel injection control to guarantee successful combustion in an individual cylinder constrained by the A/F is more difficult.

Figure 3.7 shows an example of the starting speed response, which is a simulation result obtained from a full-scale engine simulator [16]. It can be seen that the starting speed shows an extremely large overshoot, which is more than twice the required one, and therefore a long settling time, which is also more than twice the desired value. This is due to the over-torque generation during the first few engine cycles. To suppress the undesired torque generation, delicate torque management is required during the starting stage.

In this section, a starting speed control scheme that involves an individual fuel injection control law will be introduced. The starting conditions and the specified speed performance are described in Figure 3.8. The crankshaft is in a cranking state by the starting motor with an initial speed of $250 \perp 50$ rpm, and the operation time of the starting motor is no longer than 1.5 s. During the cranking stage, combustion will occur in the cylinders along the crank angle, then the engine generates torque and the starting motor is turned off. According to the absolute crank angle value, which is usually measurable after the engine acquires synchronization during the starting operation, the first combustion event in the corresponding cylinder can be identified. Moreover, there

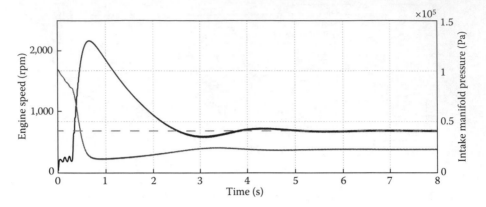

FIGURE 3.7
Response of engine speed and manifold pressure during starting with constant control inputs: both the throttle opening and the SA commands are constant; the fuel injection commands are given such that the A/F converges to the ideal value.

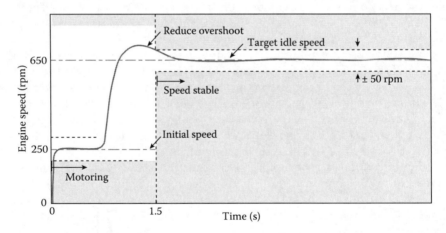

FIGURE 3.8
Image of starting speed with required specification.

should be no misfiring, and the control system is supposed to be convergent to the reference idle speed and not to cause oscillations. The design specifications for the starting speed control problem are as follows:

1. The idle speed command is 650 rpm, and the static error is in the range of ±50 rpm.

2. The settling time is no longer than 1.5 s, and the overshoot should be no more than 30%.

The input signals for the control design are the throttle opening, fuel injection, and SA, and the output is the engine speed and the signals, including the crank angle, engine speed, and air mass flow rate through the throttle, which are measurable for control design.

Challenging Points. Focusing on improving the transient performance of starting speed, the main challenges to solve the control problem include:

1. Management of individual A/F for each cylinder with sufficient precision to guarantee successful combustion condition; that is, there is no misfire.

2. Proper coordination among the multiactuators.

3. Robustness with respect to different operating conditions of the engine, especially to the cooling water temperature associated with different starting conditions.

For challenging point 1, the following fuel path control strategy is proposed.

Fuel injection is determined depending on the air charge in the cylinder. During the starting, due to the significant changes of engine characteristics and the transient response of engine speed, it is difficult to estimate the air charge. On the basis of the mean-value model of intake manifold (2.51), an observer with a slight manual parameter adjusting is designed to estimate the air charge [34, 35]. Furthermore, taking into account the fuel path dynamics, fuel injection is determined by inverse dynamics.

First, by using the model introduced in Section 2.5, the following observer is designed to estimate the air mass flow rate $\dot{m}_o$:

$$\begin{cases} \dot{\hat{p}}_m = \dfrac{RT_m}{V_m}(\dot{m}_{th} - \hat{\dot{m}}_o) \\[2ex] \hat{\dot{m}}_o = \dfrac{\rho_a V_c \eta}{4\pi p_a}\omega\hat{p}_m \end{cases} \tag{3.65}$$

where the initial condition $\hat{p}_m(0)$ is the atmospheric pressure p_a and manifold temperature T_m equals the atmospheric temperature T_a.

For obtaining the air charge of each cylinder, the estimated valve is decided discretely at each top dead center (TDC) timing. Let T_c denote the period of time occupied by one engine cycle. Then regarding the engines with six cylinders, the sampling period of the air charge estimation is $T_s = 2\pi/(3\omega)$ seconds, which in the crank angle domain is 120°.

Taking the airflow mass $\hat{\dot{m}}_o$ from the observer (3.65) at $t = lT_c$ and assuming the time of intake stroke approximately as $\hat{t}_{TDC}(l) = 2\pi/(3\omega(lT_c))$, the air charge is estimated by

$$\hat{m}_{cyl}(l) = \hat{\dot{m}}_o(lT_c) \cdot \hat{t}_{TDC}(l) \tag{3.66}$$

where l $(=1, 2, \cdots)$ denotes the TDC-based sampling index.

Furthermore, discretizing the fuel path model (2.20) with one cycle, that is, with sampling time $T_c = 4\pi/\omega$, we obtain the following fuel path model for cylinder i:

$$\begin{cases} m_{fi}(k+1) = (1 - \tau_w)m_{fi}(k) + \varepsilon_w u_{fi}(k) \\ m_{fci}(k) = \tau_w m_{fi}(k) + (1 - \varepsilon_w)u_{fi}(k) \end{cases} \qquad (3.67)$$

where $k(=1, 2, \cdots)$ denotes the cycle-based sampling index.

Hence to achieve ideal A/F, the desired fuel injection is given as

$$m_{fci}(k) = \frac{\hat{m}_{cyl}\big(6(k-1)+i\big)}{\lambda_d}, \quad (i = 1, 2, \cdots, 6) \qquad (3.68)$$

Then from the inverse dynamics of the fuel path (3.67) model, the injection command of the ith cylinder u_{fi} is generated by

$$u_{fi}(k+1) = a_i u_{fi}(k) + b_i m_{fci}(k) + c_i m_{fci}(k+1) \qquad (3.69)$$

where

$$a_i = \frac{1 - \varepsilon_w - \tau_w}{1 - \varepsilon_w}, \quad b_i = -\frac{1 - \tau_w}{1 - \varepsilon_w}, \quad c_i = \frac{1}{1 - \varepsilon_w}$$

Notice that the above proposed fuel injection control algorithm is a discontinuous system with multisampling periods and is constructed based on an observer in a continuous time domain. During the starting, the influence from the variations of the engine characteristics to the model parameters of individual cylinders is overcome by adjusting the parameters a_i, b_i, and c_i in the control algorithm (3.69). Furthermore, the two sampling indexes have the relationships $k = \text{fix}\big((l-1)/6\big) + 1$.

Finally, consider the convergency of the intake manifold observer. Error dynamics between real air intake dynamics (2.54) and the observer (3.65) is

$$\dot{\tilde{p}}_m = -\tilde{c}(\omega)\tilde{p}_m \quad \text{with} \quad \tilde{c}(\omega) = a_2\omega$$

where $\tilde{p}_m = \hat{p}_m - p_m$. Under the assumption of constant manifold temperature, the observation error $\tilde{p}_m(t)$ will converge to zero as $t \to \infty$, since $\tilde{c}(\omega) > 0, \forall \omega$.

The proposed fuel injection control algorithm is summarized in Figure 3.9.

For challenging point 2, a coordinated control strategy of SA and throttle opening is employed to achieve the speed regulation control.

A basic of speed control in the starting mode is how to manage the engine torque such that the engine is idling stably as quickly as possible. From the cranking state, the initial ignition event will cause a big acceleration of the rotational speed. Hence, if it is possible, the air charge should be reduced to suppress the torque generation. Moreover, for the torque regulation during this stage, spark timing control that has quick authority should be used.

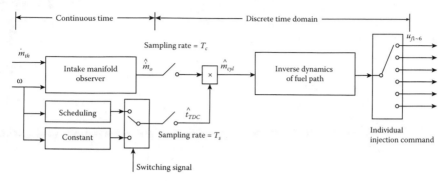

FIGURE 3.9
Structure of fuel path control system.

Based on these observations, the starting speed control loop is constructed by designing proper control laws for throttle opening and SA. For the throttle opening, the model-based feedback control presented in Section 3.3 is employed. From the initial ignition to the state that the acceleration is in a certain range, only feedback control for SA is used; then the feedback control authority is definitely switched to regulate the throttle opening. In fact, two events are specified to coordinate the feedback control authorities of SA and throttle opening, and they are defined as the following logical variables:

Event 1: $\omega(t) \geq 250$ (rpm) and $\dot{\omega}(t) \geq \Delta$

Event 2: $0 \leq e_\omega(t) \leq 50$ (rpm) and $\dot{e}_\omega(t) < 0$

where $\Delta > 0$ denotes a small constant.

First, to get the transient starting speed response more smooth, the same reference model in Section 3.3 is introduced:

$$\dot{\omega}_d(t) = -\sigma(\omega_d(t) - \omega_r), \quad t \geq t_1 \tag{3.70}$$

Here, t_1 denotes the first timing of event 1 and $\omega_d(t_1) = \omega(t_1)$. Then, the following coordinated control law for SA and throttle opening is constructed as

$$u_s(t) = \begin{cases} k_s e_\omega(t) + u_s^*, & t_1 \leq t < t_2 \\ u_s^*, & t \geq t_2 \end{cases} \tag{3.71}$$

$$u_{th}(t) = \begin{cases} 0, & t < t_2 \\ u_{th}^* + k_p(\omega)e_\omega(t), & t \geq t_2 \end{cases} \tag{3.72}$$

where k_s is the proportional gain of the SA control law, u_{s0} denotes the feedforward term, which equals the MBT, $t_2(>t_1)$ denotes the first timing of event 2, and the gain condition for $k_p(\omega)$ is deduced according to Proposition 3.1.

Figure 3.10 shows the block diagram for the above speed regulation control loop, where the trigger delivering block gives two signals according to the speed and speed error at t_1 and t_2, respectively. The first one is used to drive the reference model (3.70), and the second one determines the inputs of the feedback control loops of the throttle and SA.

The structure of the above presented starting speed control scheme can be summarized as in Figure 3.11. It includes the fuel path control loop, spark timing control loop, and throttle control loop with a supervisor for coordination management of each control loop.

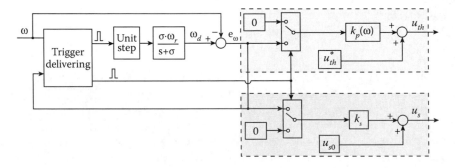

FIGURE 3.10
Control laws of SA and throttle opening.

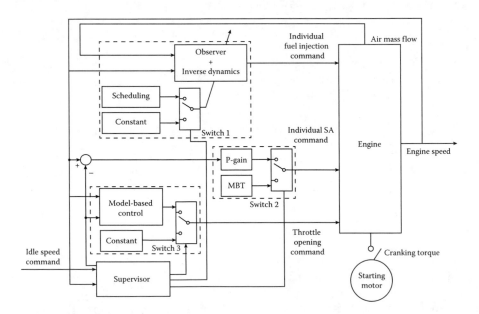

FIGURE 3.11
Block diagram of starting speed control system.

For challenging point 3, the above proposed coordinated control laws (3.71) and (3.72) for SA and the throttle opening are updated.

Regarding the starting speed control, the general uncertainty is the variation of the water temperature and the initial crank angle during each starting operation. For the robustness, the coordinated feedback control law in terms of throttle opening and SA is redesigned as follows [35]:

$$u_s(t) = k_s e_\omega(t) + u_s^*, \quad t \geq t_1 \tag{3.73}$$

$$u_{th}(t) = \begin{cases} 0, & t < t_2 \\ u_{th}^* + k_p e_\omega(t) + k_i \int_{t_2}^t e_\omega(\tau)d\tau, & t \geq t_2 \end{cases} \tag{3.74}$$

where k_p and k_i are the proportional and integral constant gains. If we compare (3.73) with (3.71), it can be observed that the SA feedback control (3.71) is turned off at $t = t_2$, while the SA feedback control (3.73) is always active for any $t \geq t_1$, and for the throttle opening, the feedback control law is redesigned using a proportional-integral (PI) controller as shown in Figure 3.12.

In the above redesigned control strategy, more authority of SA feedback control is introduced, and the integrator should be able to improve the starting speed performance in terms of reducing the settling time and eliminating the steady-state error when the operating conditions of the engine system are changed.

In the following, the convergence of the redesigned starting speed control system during the idling regulation stage, that is, $t \geq t_2$, is evaluated. The engine model used in this case for theoretical analysis is as shown below.

First, for the engine torque generation (3.4) with A/F under proper control, the linear approximation around (p_m^*, u_s^*) can be given as

$$\tau_e(t) = c_\tau p_m(t - t_d) + \left.\frac{\partial f_s}{\partial u_s}\right|_{u_s^*} \Delta u_s \tag{3.75}$$

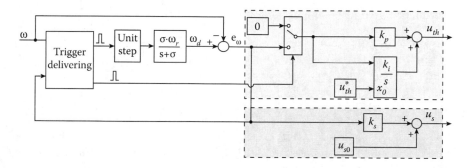

FIGURE 3.12
Updated control laws of SA and throttle opening.

where $\Delta u_s = u_s - u_s^* = k_s e_\omega$ from (3.73). Then, combining with Equation 3.75, the dynamics of engine speed in (2.59) can be rewritten as

$$\dot{\omega}(t) = a_1 p_m(t - t_d) + p_m(t - t_d)c_s k_s e_\omega(t) - \bar{D}\omega(t) - T_D \qquad (3.76)$$

with a constant c_s. Then, the nonlinear mean-value engine model (2.59) with (3.76) is linearized around the nominal state value (ω_r, p_m^*). The linearized system controlled by (3.73) and (3.74) becomes

$$\begin{cases} \dot{e}_r(t) = -\sigma e_r(t) \\ \dot{e}_\omega(t) = (\sigma - \bar{D})e_r(t) - (\bar{k}_s + \bar{D})e_\omega(t) + a_1 e_p(t - t_d) \\ \dot{e}_p(t) = -a_2 p_m^* e_r(t) - (a_2 p_m^* + a_u k_p)e_\omega(t) - a_2 \omega_r e_p(t) - a_u k_i \int_{t_2}^{t} e_\omega(\tau)d\tau \end{cases}$$
$$(3.77)$$

where $\bar{k}_s = c_s k_s p_m^*$. Define $\xi = -k_i \int_{t_2}^{t} e_\omega(\tau)d\tau$. For simplicity, suppose that the time-varying delay t_d is a constant, that is, $t_d = \pi/\omega_r$. It is clear that one can show the convergence of the speed regulation system (3.77) by proving the asymptotic stability of the closed-loop system represented by the following functional differential equation:

$$\dot{x}(t) = (A_0 + KB)x(t) + M(x(t - t_d) - x(t)) \qquad (3.78)$$

with $x = [e_\omega \ e_p \ \xi]^T$,

$$A_0 = \begin{bmatrix} 0 & a_1 & 0 \\ -a_2 p_m^* & -a_2 \omega_r & a_u \\ 0 & 0 & 0 \end{bmatrix}, \quad K = \begin{bmatrix} \bar{k}_s + \bar{D} \\ a_u k_p \\ k_i \end{bmatrix},$$

$$B = \begin{bmatrix} -1 \\ 0 \\ 0 \end{bmatrix}^T, \quad M = \begin{bmatrix} 0 & a_1 & 0 & 0 \\ 0 & 0 & 0 & 0 \\ 0 & 0 & 0 & 0 \end{bmatrix}$$

The following conclusion can be obtained.

Proposition 3.5 *For a given speed trajectory ω_d, there exists a positive-definite matrix $\Gamma \in R^{3 \times 3}$ such that the system (3.78) is asymptotically stable at the origin if the feedback gain matrix K is chosen as*

$$K = -\Gamma B^T \qquad (3.79)$$

Proof. Let $A_k = A_0 + KB$ and $G_k(s) = (sI - A_k)^{-1}M$. First, for the system (3.78), it is easy to demonstrate that (A_0, B) is detectable; then the following Riccati inequality has positive-definite solutions $\Gamma \in R^{3 \times 3}$:

$$\Gamma A_0^T + A_0 \Gamma + MM^T + \Gamma(\gamma^{-2}I - BB^T)\Gamma < 0 \qquad (3.80)$$

with a given positive constant $\gamma \leq 1$. Furthermore, one can deduce that

$$\Gamma A_k^T + A_k \Gamma + MM^T + \gamma^{-2}\Gamma^2$$
$$= \Gamma A_0^T + A_0 \Gamma + MM^T + \gamma^{-2}\Gamma^2 + \Gamma B^T K^T + KB\Gamma$$
$$\leq \Gamma A_0^T + A_0 \Gamma + MM^T + \gamma^{-2}\Gamma^2 + KB\Gamma + \Gamma B^T B\Gamma + K^T K$$

Taking the feedback gain condition (3.79) and the Riccati inequality (3.80) into account, the above inequality reduces to

$$\Gamma A_k^T + A_k \Gamma + MM^T + \gamma^{-2}\Gamma^2 < 0 \qquad (3.81)$$

which by defining $P = \Gamma^{-1}$ is equivalent to the following inequality:

$$A_k^T P + PA_k + PMM^T P + \gamma^{-2}I < 0 \qquad (3.82)$$

The Riccati inequality (3.82) has a positive-definite solution P, which implies that $G_k(s) \in RH_\infty$ and $\|G_k(s)\|_\infty \leq \gamma$. Furthermore, to analyze the stability of the time-delay system (3.78), choose a candidate of the Lyapunov–Krasovskii functional as follows:

$$V(x_t) = x^T Px + \frac{1}{\gamma^2} \int_{t-t_d}^t \|x(s)\|^2 ds \qquad (3.83)$$

where x_t represents a function defined by $x_t(\tau)$, $\tau \in [0, t_d]$, and $\|x\|^2 = x^T x$. Calculating the time derivative of (3.83) along the system (3.78) trajectory gives

$$\dot{V}(x_t) = x^T(A_k^T P + PA_k)x + 2x^T PMx(t - t_d) + \gamma^{-2}\|x\|^2 - \gamma^{-2}\|x(t - t_d)\|^2$$

Completing the square in the above equation results in the following inequality:

$$\dot{V}(x_t) \leq x^T(A_k^T P + PA_k + \gamma^2 PMM^T P + \gamma^{-2}I)x$$

Due to the condition $\gamma \leq 1$, the above inequality can be rewritten as

$$\dot{V}(x_t) \leq x^T(A_k^T P + PA_k + PMM^T P + \gamma^{-2}I)x$$

It is clear that by (3.82), there is a positive-definitive matrix $Q \in R^{3 \times 3}$ such that

$$\dot{V}(x_t) \leq -x^T Qx$$

Hence, the conclusion of Proposition 3.5 follows by the Lyapunov–Krasovskii stability theorem (B.1).

3.6   Experimental Case Study

Based on the identification result in Section 2.7, experiments are conducted to validate the proposed speed control scheme shown in Figure 3.3 and the starting speed control schemes. The validation experiments are conducted by using the same engine test bench as the one for model identification. In the experiments, the control input commands of throttle opening (in degrees), individual SAs (in degrees per cycle), and fuel injection (in milli-milliliter [mml]) per cycle U_{fi} are calculated specifically as follows:

$$\phi = \frac{180}{\pi} \cdot \arccos(1 - u_{th}) \tag{3.84}$$

$$\mathrm{SA} = \frac{180}{\pi} u_s \tag{3.85}$$

$$U_{fi} = \frac{u_{fi}}{\rho_f} \times 10^6 \tag{3.86}$$

where $\rho_f = 0.735$ (kg/L) is the density of the used gasoline.

Experimental Result of the Starting Speed Control Scheme. The engine is in a neutral state with the unengaged dynamometer. Considering that the water temperature T_w has a significant influence on the wall wetting and the combustion of the fuel in each cylinder, especially during the first firing cycle, the testing experiments are conducted for every 5°C interval in T_w within the range from almost 30°C to 75°C. Attention is also paid to the variations of the initial absolute crank angle. It may impact the air charge of the first firing cycle, and the influence from the air charge estimation combined with the associated fuel injection in terms of torque generation can cause distinct starting speed performance. In the testing, the initial crank angle of each engine starting is random, and we characterize the different experimental results corresponding to the different initial crank angles that are detected after the experiments are finished; T_w is manually selected by warming up the engine. Moreover, from a practical point of view, the SA should be limited to avoid heavy knocking. Hence, the control input of SA is restricted when the experiments are conducted, where the lower and upper constraints are set at $-10°$ and $25°$, respectively, which are decided according to the experiences in practice.

The parameter σ is selected as $\sigma = 5$. The following feedforward compensation values for throttle opening and SA are used: $\phi^* = 1.15°$, MBT $= 20°$. The nominal model parameters $\tau_w = 0.1$ and $\varepsilon_w = 0.01$ are used. The scheduling employed in the fuel injection law is a lookup table of the water temperature. The parameters of the lookup table are identified from testing experiments.

The starting speed control scheme with control laws (3.71) and (3.72) is validated first. By selecting proper control parameters in the feedback control loops and under the initial conditions, with the initial crank angle being at 540° and $T_w = 55°C$, Figure 3.13 shows a validation result, including the starting speed, the individual A/F, the given commands of throttle opening, SA, and fuel injection, as well as the air mass flow rate estimation. It can be observed that for the response of the starting speed, the overshoot is 30%,

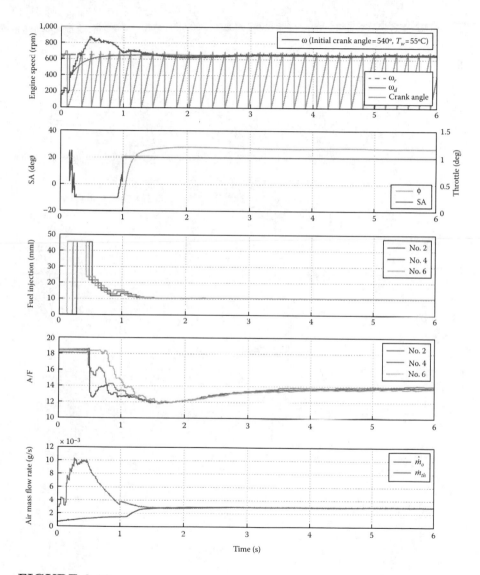

FIGURE 3.13
Experimental result of the starting speed control scheme with the control laws (3.71) and (3.72).

the settling time is 1.3 s, and the speed converges to the idling reference 650 rpm with a static speed error of less than ±20 rpm. The individual A/F is in the range that guarantees the combustion condition. It can be also seen that the output of the open-loop observer 3.65 $\hat{m}_o$ converges to the measured air mass flow rate $\dot{m}_i$. Corresponding testing experiments are conducted by applying the redesigned speed control laws (3.73) and (3.74), where the feedback gains are chosen as $k_s = 0.15$, $k_p = 5 \times 10^{-5}$, and $k_i = 5 \times 10^{-5}$. Figure 3.14 shows the experimental result.

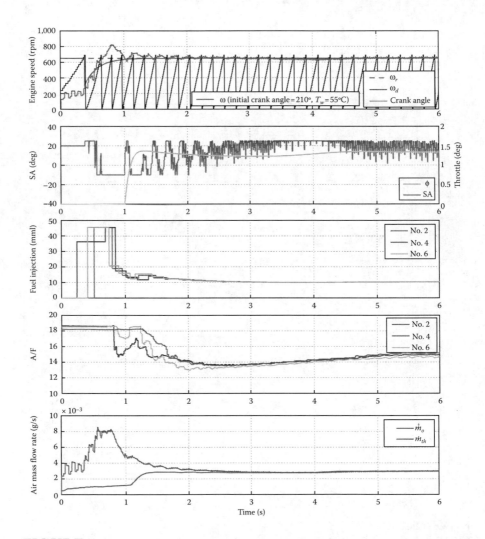

FIGURE 3.14
Experimental result of the starting speed control scheme with control laws (3.73) and (3.74).

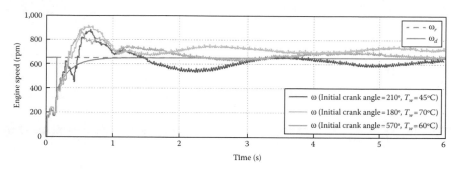

FIGURE 3.15

Speed response of the starting speed control at different initial conditions with the control laws (3.71) and (3.72).

For the robustness testing, many experiments are conducted at different initial conditions with the same control parameters as in the control laws. Figure 3.15 shows three results of the starting speed under control laws (3.71) and (3.72). The points observed from these results are illustrated as follows. There is clear low-frequency oscillation in the starting speed: this is caused by the different initial torque generation due to the different initial crank angle and T_w, as well as by the speed-dependent feedback gain in the throttle controller (3.72). Moreover, steady-state error exists when T_w is changed. It is found that the original control scheme is fairly effective only in a quite limited range of starting conditions in terms of initial crank angle and T_w. On the other hand, Figure 3.16 presents the results when using the speed control laws (3.73) and (3.74). Figure 3.16a shows the speed responses with respect to higher T_w ($>40°$C). Using the same feedforward compensation and feedback gains under lower water temperature conditions results in an undesired starting speed performance, as shown by curve I in Figure 3.16b; however, as the shown by curve II, when a larger initial condition $\phi^* = 1.75°$ of the integrator in the throttle control loop is given, the performance of the speed response is clearly improved. These results show that the starting speed response has a shorter settling time when the redesigned control scheme is applied since it involves additional feedback action to the input of SA, which has fast authority for speed regulation. However, a drawback of using a much more active SA control is to induce detrimental effects on emission. Moreover, it has been found experimentally that an effective value for k_S is in the interval $(0.05, 0.25)$. It should be pointed out that to decide the gain matrix K, the Riccati inequality (3.81) should be solved to obtain the matrix Γ. In fact, Proposition 3.5 guarantees the existence of the gain matrix K. However, using the solution of (3.81) generally cannot guarantee satisfactory performance. Hence, the gains k_p, k_i, and k_s in the final design are decided by experimental testing.

Comparing the results in Figure 3.15 and Figure 3.16a indicates that the control effort of the integrator in the feedback loop of the throttle opening

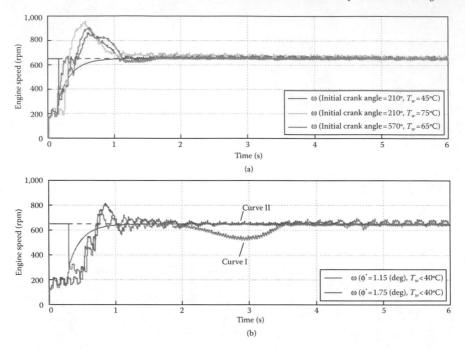

FIGURE 3.16
Speed response of the starting speed control with control laws (3.73) and
(3.74). (a) At different initial conditions; (b) at lower water temperature using
different initial conditions for the integrator.

can overcome the steady-state error. This experimental observation verifies
that the redesigned control scheme guarantees that the control system is
robust to deal with initial condition variations with more regions than the
original control scheme. On the other hand, as the result in Figure 3.16b
indicates, to guarantee the effectiveness in a larger operating range, combining
the scaling technique to the feedforward compensation according to water temperature should be considered. In other words, the proposed control scheme
is not robust enough with respect to different initial conditions. In fact, the
feedforward compensation for the throttle opening depends on the precision
of the model parameters that are varying when the operation condition of the
engine is changed. For practical application, it is clear that this feedforward
compensation is critical. To solve this issue, a systematic approach to find
proper compensation should be investigated in future work.

**Experimental Result of the Robust Adaptive Speed Control
Scheme.** The robust adaptive speed controller proposed by Proposition 3.4 is
tested. The engine operates in a speed mode at 750 rpm. The control parameters are set as $k_p = 5$ and $\rho = 0.025$. Figure 3.17 shows a testing result. Step
load disturbances with 20 and 10 Nm are applied by the dynamometer.

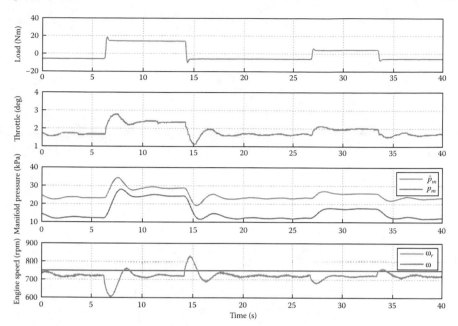

FIGURE 3.17
Speed control result under controller (3.60) with the adaptive law (3.61).

3.7 Conclusions

This chapter introduced the mean-value model-based idle speed control design approaches.

First, with explicit consideration of the intake-to-power delay, a nonlinear speed control scheme was introduced. By applying the Lyapunov–Krasovskii stability theory, a precise stability analysis of the error system was provided for the normal system. The cases using a nonlinear time-varying feedback gain and a constant one were investigated; it was also shown that under the proposed control scheme, the stability is guaranteed even though the delay time is varied according to the engine speed. The robustness of the control system to the model uncertainty was analyzed. Then, focusing on tackling the speed set-point regulation problem, this scheme was used to deal with a benchmark problem on engine starting control. The effectiveness of this control scheme was demonstrated by experimental results on the starting operating mode.

Then, with a delay compensation strategy by SA, an adaptive idle speed control scheme was introduced. This control strategy can guarantee the asymptotic stability of the desired idle speed under any load disturbance and uncertain model parameters. Moreover, the proposed throttle opening controller has a simple structure for practical applications. Experimental study demonstrates the effectiveness of the control design.

4

Air–Fuel Ratio Control

4.1   Introduction

From the foregoing chapters, it is known that the main blocks of the gasoline engine are the fuel path and the air path that define the gas mixture entering the cylinder, the combustion block that determines the amount of torque produced by the engine, and the exhaust block with the three-way catalytic (TWC) converter that can remove emissions of hydrocarbon, carbon monoxide, and nitrogen oxide. The catalytic conversion efficiency for pollutant gases significantly depends on the precise value of air–fuel ratio (A/F), since only for a very narrow A/F band around stoichiometry can all three pollutant species (HC, CO, NO_x) present in the exhaust gas be almost completely converted to the innocuous components water and carbon dioxide [3, 36, 37]. In addition to the emission concerns, regulating A/F according to the stoichiometric value can also improve the fuel economy and provide efficient torque demands. Therefore, the main goal of A/F control is to maintain the A/F close to the stoichiometric value for efficient combustion, power performance, fuel economy, and optimal after-treatment conditioning, especially during transient operations such as acceleration, deceleration, and load changes.

On the other hand, the gasoline engine is controlled by the relative air supply. The air mass flow into the manifold is controlled by the throttle opening, while the amount of fuel being mixed with the air is subsequently regulated by the injection command to the fuel injector metering the flow of fuel into the fuel path. It should be noted that the A/F is defined as the ratio of the air mass flow into the cylinder to the atomized fuel mass in the cylinder. Therefore, to control the A/F correctly at all times, especially in transients, the injected amount of fuel must be adapted to the air mass flow into the cylinder rather than into the intake manifold. However, the amount of air mass into the cylinder cannot be measured in practice; instead of direct measurement, it can be estimated. Consequently, in order to maintain the A/F close to the stoichiometric value in both steady-state and transient conditions, three main aspects are required: a correct estimation of air charge mass per inlet stroke for each cylinder, so-called cylinder air charge estimation; the fuel injection in the right amount, that is, precise fuel injection; and the closed-loop regulation of A/F utilizing the universal exhaust gas oxygen (UEGO) sensor.

First, correct cylinder air charge estimation is an important aspect for the accurate control of A/F. At a static operation mode, it is not difficult to obtain a mean-value estimation of the cylinder air charge based on the measurement of air mass flow passing through the throttle valve and the engine speed. However, at a transient mode, the challenge of cylinder air charge estimation is due to the dynamics of the intake manifold, the intake-to-power or intake-to-exhaust delay, and the sensor delay. A reasonable way to deal with this challenge is to take into account the effect of the dynamics or the delay properties in the estimation. In the past decade, several approaches have been reported that targeted the cylinder air charge estimation problem by paying attention to the dynamics of the intake manifold; for instance, see [38–45] and other relevant references. A common characteristic of most approaches is to introduce an engine dynamics-based observer into the cylinder air charge estimation algorithm. Considering the compromise of this approach on the complexity of the control algorithm and extra sensors, in this chapter, we introduce a simple estimation algorithm for cylinder air charge and present a feedforward control of A/F based on this estimation [46].

Precise fuel injection is another aspect deserving effort for the objective of A/F control. It is worth mentioning that its realization relies on the injection types—port injection, direct injection, or even dual injection—besides the correct amount of fuel required according to the cylinder air charge. The systems of direct injection and port injection show different dynamics, which leads to distinct performance during the transient mode of engine operations [47]. The essential point is the wall-wetting phenomenon in the port injection path. This indicates that when the port injection is adopted, the amount of fuel injection should involve the compensation amount increased or reduced for the wall-wetting phenomenon. For engines with port injection systems, a lot of attention has been paid to the modeling of the fuel path dynamics, and it is demonstrated that the first-order model proposed in [13] is effective to characterize the wall-wetting phenomenon. Moreover, the identification of this fuel path model has been investigated significantly for engines with distinct structures and for engines operating under different conditions [48–51]. Using a dual-injection system is an effective way to more precisely mix air and fuel and could achieve greater efficiency [52–54]. In this kind of engine, the fuel injection is switched or coordinated between the two paths according to the engine operating mode. Accordingly, it will result in the potential of each injection method to be used by promptly choosing the injection path that meets the real-time operation environment. For engines with dual injection, besides paying attention to the fuel path modeling and identification, focusing on the increased or decreased compensation amount for the transient process of injection path switching is essential to precise fuel injection. Therefore, in this chapter we mainly present a wall-wetting model-based control scheme for A/F transient control of engines with a dual-injection system [55].

For achieving the accurate regulation of stoichiometric A/F, the closed-loop control utilizing the feedback signal of the UEGO sensor is an

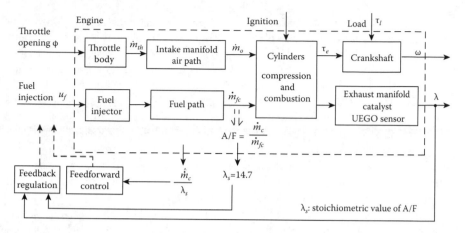

FIGURE 4.1
Schematic of A/F control system of gasoline engines.

essential part. Figure 4.1 shows the schematic of the A/F control system of gasoline engines. The feedback control is especially important when there exist inaccuracies in the air charge estimation and the wall-wetting compensation resulting from the wide engine operating range, the inherent nonlinearities of the combustion process, the large modeling uncertainties, and parameter variations. The uncertainty of plant parameters, depending on the operation mode, motivated the introduction of adaptive control algorithms into the fuel injection controller. Here typical examples are the Kalman filter-based adaptive estimator for fuel film dynamics [56], sliding mode-based adaptive control for the cylinder airflow and fueling parameters [57], and adaptive internal model control [58], adaptive posicast control [59], and adaptive reference model control with Smith predictor [60] for sensor aging and transmission time delay. However, in this chapter, we will present an adaptive A/F control strategy containing the feedforward compensation and feedback regulation with the adaptive update law [61], which is based on a mean-value engine model that accounts for the impingement of the injected fuel on the walls and the evaporation process with uncertainty and calculates the uncertain parameters in the dynamics of the intake manifold and crankshaft rotation.

In addition, due to the appearance of the terminologies of feedforward control and adaptive control herein and before, and the use of their design concepts in the A/F control system, their basic idea and principle will be described simply as follows.

The basic principle of feedforward control is to adjust the control input according to the measured perturbation, including the variation of the set-point input and the external disturbance, such that the effect of the measured perturbation on the process output is reduced or even eliminated to improve the control accuracy of the system output. The difference between feedforward and feedback lies in that feedforward control is based on knowledge about the measured perturbation, instead of error between the set-point value and the real value of the controlled variable. The measured signal in a feedforward system is the variation of perturbation rather than the controlled variable, and the control action in a feedforward system generates instantly following the perturbation action, without the need to wait until after the deviation occurs. However, feedforward control is always used along with feedback control because a feedback control system is required to track set-point changes and suppress unmeasured disturbances, while combined feedforward plus feedback control can significantly improve performance over simple feedback control whenever there is a major disturbance that can be measured before it affects the process output. Figure 4.2 gives the traditional block diagrams of two kinds of feedforward control systems.

Adaptive control is a control method used for a controller that must adapt to a system with unknown or time-varying parameters, so-called parameter uncertainty. It is well known that feedback is basically used in conventional control systems to reject the effect of disturbances upon the controlled variables and to bring them back to their desired values according to a certain performance index. To achieve this, the measurements of the controlled variables are compared with their desired values and the differences are fed into the controller, which will generate the appropriate control. In fact, a similar conceptual approach can be considered for solving the problem of achieving and maintaining the desired performance of a control system with parameter uncertainty. The calculated performance utilizing the inputs, the states, the outputs, and the known disturbances for the control system is compared to the desired performance, and their differences (if the real-time performance is not acceptable) will be fed into an adaptive mechanism. The adaptive mechanism modifies parameters of an adjustable controller or generates an auxiliary control in order to achieve the system performance accordingly within the set of acceptable ones. The adaptive mechanism and the adjustable controller constitute a complete adaptive controller. A block diagram illustrating a basic configuration of an adaptive control system is given in Figure 4.3.

When designing adaptive control systems, special consideration of convergence and robustness issues is necessary. Lyapunov stability is typically used to derive control adaptation laws, and projection (mathematics) and normalization are commonly used to improve the robustness of estimation

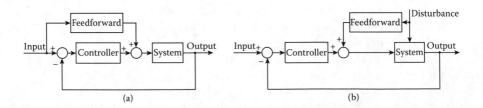

FIGURE 4.2
Feedforward control system.

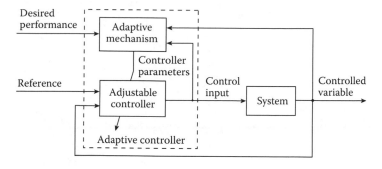

FIGURE 4.3
Basic configuration for an adaptive control system.

algorithms. The details of Lyapunov stability and the design procedure of adaptive controller can be found in Appendix A.

4.2 Air Charge Prediction-Based Feedforward Control

In this section, we are concerned with the cylinder air charge estimation-based A/F control for direct injection engines with more than four cylinders during the transient operation. Here suppose that a combustion is complete and the residual gas can be ignored. Figure 4.4 shows the sketch of a gasoline engine with six cylinders sharing a common manifold.

Specifically, in Figure 4.4 cylinders 1 to 6 indicate the order of the firing sequence. p_m denotes the manifold pressure, $\dot{m}_{th}$ represents the air mass flow rate passing through the throttle valve, and $\dot{m}_{c_i}$ $(i = 1, 2, \cdots, 6)$ is the air mass flow rate leaving from the manifold and passing through the intake valve of the ith cylinder. The objective is to estimate the air mass charged into the individual cylinder during transient operation.

From this sketch, it can be seen during transient operation that the air mass flow passing through the throttle valve is not equivalent to the air mass flow leaving the manifold, which results from the variation of the manifold pressure, furthermore, the air mass flow leaving the manifold will be distributed by no less than two overlapping cylinders in a certain interval of crank angle domain.

Based on these observations, an estimation for the individual cylinder air charge mass will be given as follows.

Suppose that the intake valve of each cylinder opens at the corresponding top dead center (TDC) and closes at the bottom dead center (BDC). As is well known, in the continuous time domain the air dynamics in the intake manifold is represented as follows:

$$\dot{p}_m = \frac{RT_m}{V_m}(\dot{m}_{th} - \dot{m}_o) \tag{4.1}$$

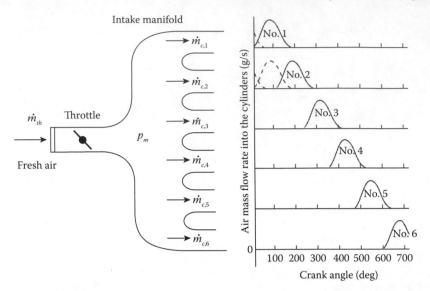

FIGURE 4.4
Sketch of the air mass flow rate into cylinders.

For cylinder i, let $t_{TDC,i}$ and $t_{BDC,i}$ denote the timing of the corresponding TDC and BDC, respectively. Then, during the intake stroke of cylinder i, that is, in the time domain $[t_{TDC,i}, t_{BDC,i}]$, we have

$$
\dot{m}_o = \begin{cases} \dot{m}_{c,i-1} + \dot{m}_{c,i}, & t_{TDC,i} \leq t < t_{TDC,i+60} \\ \dot{m}_{c,i}, & t_{TDC,i+60} \leq t < t_{TDC,i+1} \\ \dot{m}_{c,i} + \dot{m}_{c,i+1}, & t_{TDC,i+1} \leq t \leq t_{BDC,i} \end{cases} \tag{4.2}
$$

This means that the total air mass flow rate leaving the manifold will be distributed to two cylinders, and the cylinder-to-cylinder relays with a 60° interval in the crank angle domain along the firing sequence. Based on this observation, the ith cylinder air charge $m_{c,i}$ can be approximately represented as

$$
m_{c,i} = \int_{TDC,i}^{TDC,i+180} \dot{m}_{c,i}(t)dt = \int_{TDC,i}^{TDC,i+120} \dot{m}_o(t)dt \tag{4.3}
$$

Assign the sampling time to TDC timing and indicate the sampling index by $k = 0, 1, 2, \cdots$; that is, the sampling period T_s is 120° in the crank angle domain. Then at the kth sampling step, $m_c(k)$ denotes the total air mass charged into the individual cylinder, in which the intake stroke proceeded at 120°. And from (4.1), it follows that

$$
m_c(k) = m_{th}(k) - \frac{V_m}{RT_m} \{ p_m(k) - p_m(k-1) \} \tag{4.4}
$$

where

$$m_{th}(k) = \int_{(k-1)T_s}^{kT_s} \dot{m}_{th}(t)dt$$

Therefore, the formulation (4.4) can be regarded as an estimation for the air mass flow charged into the individual cylinder. Its physical meaning is that at a static mode, the amount of air mass passing through the throttle valve $m_{th}(k)$ equals the cylinder air charge, since $p_m(k) = p_m(k-1)$. Meanwhile, at a transient mode the difference of manifold pressure effects the cylinder air charge.

However, it should be noted that the accuracy of the estimation (4.4) depends on the parameter of the model, and the parameter value usually fluctuates when the thermal condition changes. To compensate the modeling error and uncertainties, an adjusting law based on the A/F measurement should be introduced into the estimation. Nevertheless, synchronization of the detected signals must also be considered, since the A/F of a combustion in a cylinder can be measured with the compression exhaust and sensor delay. Accordingly, Figure 4.5 shows the diagram of the air charge estimation algorithm with consideration of the delay in the path from fueling to the sensor output.

In Figure 4.5, $u_f(k)$ is the direct fuel injection command, d_1 denotes the sampling period delay from the fuel injection command $u_f(k)$ to the implementation of fueling $m_{fc}(k)$, and d_2 denotes the delay steps from the end of the fueling implementation to the beginning of the exhaust stroke; furthermore, $d = d_1 + d_2$ represents the delay steps from the command to the beginning of the exhaust stroke, and $G(z)$ denotes the transfer function of the exhaust and

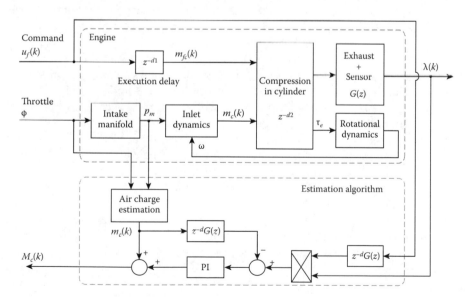

FIGURE 4.5
Diagram of the air charge estimation algorithm.

the sensor, which is usually modeled as a first-order lag with a certain time constant.

From Figure 4.5, the estimation error detected at the timing of the sensor output is

$$e(k) = \lambda(k)z^{-d}G(z)u_f(k) - z^{-d}G(z)m_c(k) \qquad (4.5)$$

Using this error, we introduce a real-time adjusting term into the estimation model (4.4). Consequently, the individual cylinder air charge estimation with the adjusting law is given as follows:

$$M_c(k) = m_c(k) + k_P e(k) + k_I \sum_{i=0}^{k} e(i) \qquad (4.6)$$

where k_P and k_I are gains, and the last term is to compensate the unknown static offset.

Based on the proposed estimation algorithm, the A/F control law can be provided for the direct injection engines at transient operation mode.

With the help of the air charge estimation (4.6), the fuel injection command $u_f(k)$ for the individual cylinder can be determined directly by the following feedforward form:

$$u_f(k) = \frac{1}{\lambda_s} \left\{ m_c(k) + k_P e(k) + k_I \sum_{i=0}^{k} e(i) \right\} \qquad (4.7)$$

However, as shown in Figure 4.5, it should be noted that there is a d_1 sampling period delay from the fuel injection command $u_f(k)$ to the implementation of fueling $m_{fc}(k)$. It results from the communication delay between the electrical control unit (ECU) and the dSPACE rapid prototyping unit implementing the proposed controller, and the time required for ECU to calculate the injected fuel mass and write the fuel command to memory before the intake stroke. Namely, the current fuel injection command $u_f(k)$ is implemented after the d_1 sampling period. Consequently, the individual cylinder air charge mass must be predicted d_1 steps ahead. Meanwhile, it can be seen from (4.4) that the prediction for the individual cylinder air charge estimation $\hat{m}_c(k + d_1|k)$ in fact is the prediction for the intake manifold pressure $\hat{p}_m(k + d_1|k)$. Thereby, the prediction algorithm for the intake manifold pressure is given as follows.

First, for calculating the prediction of the intake manifold pressure $\hat{p}_m(k + d_1|k)$, the model (4.1) should be transformed into the discrete form, and also, the mean-value model $\dot{m}_o = \frac{V_c \eta}{120RT_m} p_m n_e$ is considered; n_e denotes the engine speed in rpm units, $n_e = 30\omega/\pi$. In addition, it should be noted that the model is discrete in the crank angle domain with a 120° sampling period. As a result, the intake manifold pressure model is described as

$$p_m(k+1) = (1 - a_2)p_m(k) + a_1 \frac{\dot{m}_{th}(k)}{n_e(k)} + \varepsilon(k) \qquad (4.8)$$

$$\text{with } a_1 = \frac{30}{\pi} \cdot \frac{RT_m}{V_m}, \quad a_2 = \frac{V_c \eta}{4\pi V_m}$$

where $\varepsilon(k)$ denotes white noise with zero mean and variance σ^2. The model (4.8) can be regarded as the one-step head prediction model that is true of the minimum-variance prediction. Nevertheless, here consider $d_1 = 2$ and assume that the air mass flow rate $\dot{m}_{th}(k)$ passing through the throttle valve remains in the following two sampling periods.

Subsequently, on the basis of the minimum-variance prediction,

$$\min\left\{E\left[\sum_{i=1}^{2}[\hat{p}_m(k+i|k) - p_m(k+i)]^2\right]\right\} \tag{4.9}$$

the prediction of intake manifold pressures $\hat{p}_m(k+1|k)$ and $\hat{p}_m(k+2|k)$ can be calculated by the following iterative form:

$$\begin{cases} \hat{p}_m(k+1|k) = (1-\hat{a}_2)p_m(k) + \hat{a}_1\dfrac{\dot{m}_{th}(k)}{n_e(k)} \\ \hat{p}_m(k+2|k) = (1-\hat{a}_2)^2 p_m(k) + \hat{a}_1(2-\hat{a}_2)\dfrac{\dot{m}_{th}(k)}{n_e(k)} \end{cases} \tag{4.10}$$

where $\hat{a}_1, \hat{a}_2$ are the identification for a_1, a_2, which are obtained by the on-line recursive least-squares algorithm.

Therefore, based on the prediction for intake manifold pressure $\hat{p}_m$ $(k+1|k)$, $\hat{p}_m(k+2|k)$, the two-step $(d_1 = 2)$ prediction of the air charge estimation can be obtained as follows:

$$\hat{m}_c(k+2|k) = \alpha_0\frac{\dot{m}_{th}(k)}{n_e(k)} + \sum_{j=0}^{1}\beta_j\hat{p}_m(k-j+2|k) \tag{4.11}$$

$$\text{with } \alpha_0 = 120, \quad \beta_0 = -\frac{V_m}{RT_m}, \quad \beta_1 = \frac{V_m}{RT_m}$$

Furthermore, the fuel injection command $u_f(k)$ can be given by the following air charge prediction-based feedforward control:

$$u_f(k) = \frac{1}{\lambda_s}\left\{\hat{m}_c(k+2|k) + k_P e(k) + k_I \sum_{i=0}^{k} e(i)\right\} \tag{4.12}$$

4.3   Wall-Wetting Model-Based Feedforward Control

In this section, we focus on the aspect of precise fuel injection for the objective of A/F control. We mainly present a wall-wetting model-based control scheme for A/F transient control of V6-type gasoline engines with a dual-injection system. Figure 4.6 shows the structure of this kind of A/F control system, where a bank consisting of three cylinders is sketched. For each cylinder in the engine system, there are two fuel injectors equipped at the inlet port near

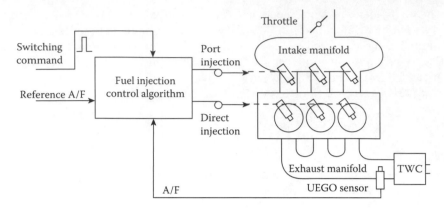

FIGURE 4.6

Sketch of fuel injection control system.

the intake valve and the inside of the cylinder, respectively: the latter is used to directly inject the fuel into the cylinder, and the former injects the fuel into the inlet port to supply into the cylinder as premixed gas. Three cylinders of the six cylinders construed in a bank share a common exhaust manifold, and a UEGO sensor for measuring the A/F λ is equipped at the gas mixing point of each exhaust manifold.

The direct injection and the port injection show different fuel dynamics, and the essential point is the wall-wetting phenomenon in the port injection path. As introduced in Chapter 2, for port fuel injection, the liquid fuel injected into the intake port only partially enters the cylinder to participate in the immediate combustion event. A portion is stored in fuel puddles at the intake port walls and at the back face of the intake valve, to be subsequently inducted into the cylinder for later combustion events. This is the so-called wall-wetting phenomenon. This indicates that when the authority of port injection is increased or reduced, using constant port injection for steady-state operation, the balancing of the fuel path will be broken during the transient stage of the port injection path, and consequently, the A/F will leave the desired value. Figure 4.7 shows the transient A/F response without consideration of the wall-wetting behavior when the injection path is switched from the port injection to the direct injection and the inverse case. In Figure 4.7 t_{Sk} denotes the switching timing.

It can be observed that when the fuel injection command is changed from the port injector to the direct injector with a constant value for the desired A/F λ_s, the A/F will be rich since the fuel remaining in the port is injected during a short period. Conversely, it will be lean when the switching signal is triggered. Therefore, in the following a control scheme will be presented for improving the performance of A/F control during the transient stage as the authority of port injection command is changed.

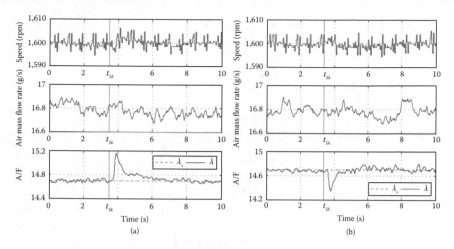

FIGURE 4.7

Response of A/F during injection path changing. (a) Direct injection switches to port injection; (b) port injection switches to direct injection.

To this end, suppose that the switching between the dual-injection paths occurs at a static mode and the total fuel injection command u_f^* for obtaining the desired A/F λ_s is known. Meanwhile, denote g_s as the index of the port injection command authority, $g_s \in [0, 1]$, and at a switching timing t_{Sk}, $g_s(t_{Sk}^-) < g_s(t_{Sk}^+)$ means that the authority of port injection is increased; otherwise, it means that the authority of port injection is reduced. Accordingly, a feedforward dynamic compensation is deduced according to the following description.

The wall-wetting phenomenon of the port injection path dynamics is described as (2.20) stated in Section 2.3.2, which is rewritten as the following:

$$\begin{cases} \dot{m}_{fw} = -\tau_w m_{fw} + \varepsilon_w u_{fp} \\ m_{fc} = \tau_w m_{fw} + (1 - \varepsilon_w) u_{fp} \end{cases} \qquad (4.13)$$

where u_{fp} denotes the port injection command.

Now, consider the dynamics of the fuel entering into the cylinder after the switching timing t_{Sk} when the injection mode is changing.

First, consider the case of changing the port injection mode to the direct injection mode at the timing t_{Sk}, that is, with $g_s(t_{Sk}^-) = 1$ and $g_s(t_{Sk}^+) = 0$.

Then, at $t = t_{Sk}$, the fuel puddle mass on the wall is

$$\tau_w m_{fw}(t_{Sk}) = \varepsilon_w u_{fp}(t_{Sk}^-) = \varepsilon_w u_f^*(t_{Sk}^-)$$

which enters the cylinder gradually through evaporation after t_{Sk}.

Consequently, after the timing t_{Sk}, the total fuel mass entering into the cylinder contains two parts. One part is the injected fuel mass

$m_{fc1}(t) = u_{fd} = u_f^*$, where u_{fd} denotes the direct injection command. The other can be represented as an unforced response with an initial value of the state $m_{fw}(t_{Sk})$:

$$m_{fc2}(t) = \varepsilon_w u_{fp}(t_{Sk}^-)e^{-\tau_w(t-t_{Sk})}$$

$$= \varepsilon_w u_f^*(t_{Sk}^-)e^{-\tau_w(t-t_{Sk})}, \qquad t \geq t_{Sk} \qquad (4.14)$$

And then, consider the converse case; that is, the injection path is switched from the direct injection to the port injection. The fuel mass injected directly will be cut at the switching time t_{Sk}; however, the fuel mass delivered from the port injection will be determined as follows: one part is a fraction $1 - \varepsilon_w$ of the injected fuel mass $u_{fp}(t_{Sk})$ entering into the cylinder immediately, and the other is the forced response of the fuel puddle resulting from $\varepsilon_w u_{fp}(t_{Sk})$, with an initial value of the state $m_{fw}(t_{Sk}^-) = 0$:

$$\tau_w m_{fw}(t) = \varepsilon_w u_{fp}(t_{Sk}) \left(1 - e^{-\tau_w(t-t_{Sk})}\right), \quad t \geq t_{Sk}$$

That is, the total fuel mass entering into the cylinder by the port injection is determined as

$$m_{fc}(t) = u_{fp}(t_{Sk}) - \varepsilon_w u_{fp}(t_{Sk})e^{-\tau_w(t-t_{Sk})}, \quad t \geq t_{Sk} \qquad (4.15)$$

According to the above analysis derivation, it is evident that the following dynamic feedforward compensation, which works in the direct injection path,

$$\alpha(\tau_w, \varepsilon_w) = -\text{mod}_k \cdot \varepsilon_w u_f^*(t_{Sk}^-)e^{-\tau_w(t-t_{Sk})}, \quad t \geq t_{Sk} \qquad (4.16)$$

with the mode switching mod_k defined as

$$\text{mod}_k = \begin{cases} -(1 - g_s(t_{Sk}^-)), & g_s(t_{Sk}^-) < g_s(t_{Sk}^+) \\ g_s(t_{Sk}^-), & g_s(t_{Sk}^-) > g_s(t_{Sk}^+) \end{cases} \qquad (4.17)$$

can result in $u_f \rightarrow u_f^*$ during the transient injection path switching mode. $u_f = u_{fd} + u_{fp}$ denotes the total fuel injection command of the engine system.

Then, after a mode switching occurs at t_{Sk}, the fuel injection commands delivered to direct injection and port injection are as follows, respectively:

$$\begin{cases} u_{fd} = (1 - g_s)u_f^* + \alpha(\tau_w, \varepsilon_w) \\ u_{fp} = g_s u_f^* \end{cases} \qquad (4.18)$$

So far it gives a dynamic feedforward compensation law to the direct injection path based on the wall-wetting model of the port injection path. The block diagram of the control scheme with the presented dynamic feedforward compensation law is as shown in Figure 4.8, where the switching action is synchronized to the impulse signal.

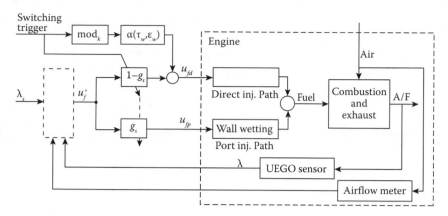

FIGURE 4.8
Block diagram of the proposed control scheme.

Remark 4.1   The signal $\varepsilon_w u_f^*(t_{Sk}^-)e^{-\tau_w(t-t_{Sk})}$ can be generated by the first-order linear transfer function with the impulse excitation $\delta(t-t_{Sk})$ at the switching time t_{Sk}.

Remark 4.2   As shown in the Figure 4.8 control scheme, the block for generating u_f^* is regarded as the design in an independent loop with the measured A/F λ and airflow.

Remark 4.3   It should be noted that knowing the model parameters τ_w and ε_w is essential to evaluate the compensator (4.16). However, these parameters are dependent on the engine operation conditions. Hence, τ_w and ε_w will be obtained by an experimental identification method with respect to the corresponding different engine operation conditions.

4.4   Lyapunov-Based Adaptive Control

In this section, aiming at inaccuracies in the air charge estimation and the wall-wetting compensation resulting from the wide engine operating range, the inherent nonlinearities of the combustion process, the large modeling uncertainties, and the parameter variations, an adaptive fuel injection control strategy will be presented to achieve the foregoing A/F control objective regardless of the various uncertainties of the system description.

   Control-Oriented System Model. First, it should be noted that the A/F control scheme design given in this section is based on the mean-value model (details can be seen in Section 2.5), suitable for the design of the model-based engine control strategies. Accordingly, here the system description used in the design includes three parts, for the convenience of adaptive control design, which are rewritten as follows.

One part is the airflow dynamics through the intake manifold, as the mean-value model (2.54) stated in Section 2.5, which can be represented as

$$\dot{p}_m = \frac{RT_m}{V_m}(\dot{m}_{th} - c_m p_m \omega) \quad \text{with} \quad c_m := \frac{\rho_a V_c \eta}{4\pi p_a} \tag{4.19}$$

Another part is the fuel delivery dynamics into the cylinder, which is described as (2.20) stated in Section 2.3.2, but here is rewritten in terms of the fuel mass flow rate as follows:

$$\begin{cases} \dot{m}_{fc} = \tau_w \dot{m}_{fw} + (1 - \varepsilon_w)u_f \\ \ddot{m}_{fw} = -\tau_w \dot{m}_{fw} + \varepsilon_w u_f \end{cases} \tag{4.20}$$

where u_f is the fuel injection command as the commanded fuel mass flow rate.

The third part is the crankshaft rotational dynamics for torque production, as introduced in Sections 2.5 and 3.2, which can be described as follows by combining the mean-value engine torque generation, (2.57) and (3.4), and considering the control law for spark advance u_s as (3.51).

$$J\dot{\omega} = c_\tau p_m - D\omega - \tau_l \quad \text{with} \quad c_\tau := \frac{\rho_a c_f \eta Q V_c}{4\pi p_a} \tag{4.21}$$

Moreover, for the adaptive control design, here the A/F is defined by $\lambda := \dot{m}_o / \dot{m}_{fc}$, which is based on the air and fuel mass flow rate with the cycle-based mean-value model. Accordingly, $\lambda = c_m p_m \omega / \dot{m}_{fc}$, a function on the system states $p_m, \omega, \dot{m}_{fc}$, can be regarded as the system output variable.

Therefore, the control-oriented system model consisting of the subsystems (4.19) through (4.21) can be rewritten as follows:

$$\begin{cases} \dot{p}_m = \frac{RT_m}{V_m}(\dot{m}_{th} - c_m p_m \omega) \\ J\dot{\omega} = c_\tau p_m - D\omega - \tau_l \\ \ddot{m}_{fc} = -\tau_w \dot{m}_{fc} + (1 - \varepsilon_w)\dot{u}_f + \tau_w u_f \\ y = \lambda = \frac{c_m p_m \omega}{\dot{m}_{fc}} \end{cases} \tag{4.22}$$

where the third equation is obtained by differentiating the first equation of (4.20) with respect to time and combining with the second equation in (4.20). $p_m, \omega, \dot{m}_{fc}$ are system state variables, u_f is the control input signal, and y is the controlled output. The system parameters c_m, c_τ, ε_w, τ_w may hold uncertainties.

In addition, it is mentioned that the measurable signals are the air mass flow rate at the throttle $\dot{m}_{th}$, the manifold pressure p_m, the engine speed ω, and the A/F λ, but the total fuel flow rate $\dot{m}_{fc}$ is not measurable. Moreover, there exists the delay between fuel injection and UEGO sensor measurement, which may limit the achievable performance of the A/F control loop.

Based on the system model (4.22), the fueling controller for achieving A/F regulation will be designed in the following. Meanwhile, the inputs to the A/F controller generally utilize the measurable signals, including intake air mass flow $\dot{m}_{th}$, engine speed ω, intake manifold pressure p_m, load torque τ_l, and predefined A/F stoichiometric value λ_s, as well as a feedback signal λ from the UEGO sensor. The output signal from the A/F controller is the required amount of fuel injection in terms of the fuel mass flow rate, which is then used to control the fuel injector metering the flow of fuel into the intake system.

Fueling Control with Feedback Compensation. First, consider the case where the system parameters c_m, c_τ, ε_w, τ_w in the system (4.22) are known. Then, the control objective is to regulate the A/F close to the stoichiometric value in the presence of the variations in engine speed and manifold pressure, and the perturbation in A/F. Accordingly, the control law can be developed as follows.

The total injected fuel amount is determined by

$$u = u_{ff} + u_{fb} \tag{4.23}$$

where u_{ff} represents the feedforward component providing the necessary amount of fuel, which is directly calculated by the steady-state airflow rate out of the manifold. u_{fb} represents the feedback components, which provides the compensative amount of fuel for the perturbations in engine speed, manifold pressure, and A/F:

$$\begin{cases} u_{ff} = \dfrac{\tau_w \dot{m}_o}{\lambda_s} = \dfrac{\tau_w c_m p_m \omega}{\lambda_s} \\[3mm] u_{fb} = \dfrac{\dot{m}_o}{\lambda} + k\tilde{\lambda} \\[3mm] \quad = \dfrac{RT_m}{V_m} \cdot \dfrac{c_m}{\lambda} \omega (\dot{m}_{th} - c_m p_m \omega) + \dfrac{c_m p_m}{J\lambda}(c_\tau p_m - D\omega - \tau_l) + k\tilde{\lambda} \end{cases} \tag{4.24}$$

where $k > 0$ is a tuning parameter and $\tilde{\lambda} = \lambda - \lambda_s$ represents the A/F steady-state error.

Furthermore, by taking the wall-wetting effect into consideration, the fueling command u_f is given by the total fuel amount with a first-order filter:

$$u_f = \frac{1}{(1 - \varepsilon_w)s + \tau_w} u \tag{4.25}$$

The convergence and stability analysis for the fueling control system is as follows.

Define a positive-definite function as

$$V = \frac{1}{2}(\lambda - \lambda_s)^2 \tag{4.26}$$

Then, differentiating V yields

$$\dot{V} = (\lambda - \lambda_s)\left\{\frac{\ddot{m}_o}{\dot{m}_{fc}} - \frac{\dot{m}_o}{\dot{m}_{fc}^2}\ddot{m}_{fc}\right\} \qquad (4.27)$$

Considering the equation about $\ddot{m}_{fc}$ and the A/F definition λ of the system (4.22) yields

$$\dot{V} = \tilde{\lambda}\frac{\dot{m}_o}{\dot{m}_{fc}^2}\left\{\frac{\ddot{m}_o}{\lambda} + \tau_w\frac{\dot{m}_o}{\lambda} - (1 - \varepsilon_w)\dot{u}_f - \tau_w u_f\right\} \qquad (4.28)$$

with $\tilde{\lambda} = \lambda - \lambda_s$. When you choose the control input signal as (4.25) with (4.23), that is,

$$(1 - \varepsilon_w)\dot{u}_f + \tau_w u_f = \frac{\tau_w \dot{m}_o}{\lambda_s} + \frac{\ddot{m}_o}{\lambda} + k\tilde{\lambda} \qquad (4.29)$$

it gives

$$\dot{V} = -k\frac{\dot{m}_o}{\dot{m}_{fc}^2}\tilde{\lambda}^2 - \frac{\tau_w\lambda}{\lambda_s}\tilde{\lambda}^2 \le 0 \qquad (4.30)$$

Therefore, according to the Lyapunov stability theorem, it follows from (4.26) and (4.30) that the A/F λ can be maintained to converge to the stoichiometry ratio λ_s.

Remark 4.4 From the fuel injection controller (4.25) with (4.23), (4.24), it is noted that in fact the controller is a state feedback control with state variables p_m, ω, λ. The fueling input consists of the direct computation $\tau_w\dot{m}_o/\lambda_s$ using the measured engine speed and manifold pressure, the fuel compensation required for the corresponding variation in the airflow rate out of the manifold $\dot{m}_o/\lambda$, and the regulation of the A/F absolute steady-state error $k\tilde{\lambda}$.

Fueling Control with Adaptive Compensation. Now, the inaccuracies of airflow and fuel delivery into the cylinders and engine torque production are under consideration, namely, the system parameters c_m, c_τ, ε_w, τ_w are regarded as the uncertain parameters, to design an adaptive fuel injection controller u_f such that the resulting closed-loop system possesses the stability and satisfactory A/F during both static and transient engine operation modes regardless of the uncertain parameters, disturbance, and time delay. Figure 4.9 shows a schematic overview of the basics of the designed adaptive controller.

The design procedure is shown as follows. It can be seen that the controller (4.25) with (4.23), (4.24), can be rewritten as

$$\frac{1 - \varepsilon_w}{c_m}\dot{u}_f + \frac{\tau_w}{c_m}u_f = \tau_w\frac{p_m\omega}{\lambda_s} + \frac{RT_m}{V_m}\frac{\omega}{\lambda}(\dot{m}_{th} - c_m p_m\omega)$$

$$+ \frac{p_m}{J\lambda}(c_\tau p_m - D\omega - \tau_l) + k\tilde{\lambda} \qquad (4.31)$$

where $k > 0$ is a tuning parameter. The uncertain parameters $(1 - \varepsilon_w)/c_m$, τ_w/c_m, τ_w, c_m, c_τ in the fuel injection controller (4.31) are replaced with

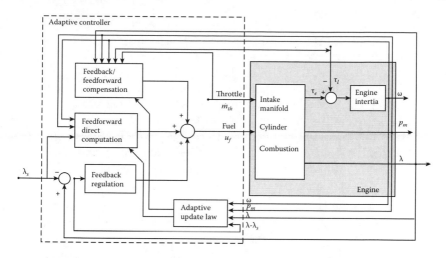

FIGURE 4.9
Schematic diagram of the A/F adaptive controller.

their estimates produced by the designed adaptive update law; consequently, it follows the proposition.

Proposition 4.1 *For the system (4.22), an adaptive fuel injection controller is given by*

$$\hat{\theta}_1 \dot{u}_f + \hat{\theta}_2 u_f = \hat{\theta}_3 \frac{p_m \omega}{\lambda_s} + \frac{RT_m}{V_m} \frac{\omega}{\lambda} (\dot{m}_{th} - \hat{\theta}_4 p_m \omega) + \frac{p_m}{J\lambda} (\hat{\theta}_5 p_m - D\omega - \tau_l) + k\tilde{\lambda}$$

$$(4.32)$$

with the parameter adaptive update laws as follows:

$$
\begin{cases}
\dot{\hat{\theta}}_1 = -\gamma_1 \dfrac{\lambda^2}{p_m \omega} \dot{u}_f \tilde{\lambda} \\[2mm]
\dot{\hat{\theta}}_2 = -\gamma_2 \dfrac{\lambda^2}{p_m \omega} u_f \tilde{\lambda} \\[2mm]
\dot{\hat{\theta}}_3 = \gamma_3 \dfrac{\lambda^2}{\lambda_s} \tilde{\lambda} \\[2mm]
\dot{\hat{\theta}}_4 = -\gamma_4 \dfrac{RT_m}{V_m} \omega \lambda \tilde{\lambda} \\[2mm]
\dot{\hat{\theta}}_5 = \gamma_5 \dfrac{p_m}{J\omega} \lambda \tilde{\lambda}
\end{cases}
\qquad (4.33)
$$

where $\hat{\theta}_i$ are estimates of the uncertain parameters θ_i, $i = 1, 2, \cdots, 5$, with $\theta_1 = (1 - \varepsilon_w)/c_m$, $\theta_2 = \tau_w/c_m$, $\theta_3 = \tau_w$, $\theta_4 = c_m$, $\theta_5 = c_\tau$, γ_i $(i = 1, 2, \cdots, 5) > 0$ as the tuning parameters. Then, the fueling control can regulate the A/F close to the stoichiometric value in the presence of the uncertainties in fuel delivery dynamics and the measurement bias of the air mass flow rate into the manifold.

Remark 4.5 It can be noted that the fueling control input (4.32) is still made up of the direct computation term (the first term), the fuel compensation term with the oxygen sensor feedback (the second and third terms), and the regulation term of the A/F (the last term); only the corresponding parameter estimations with the adaptive update laws are utilized instead of the uncertain parameters. The fuel injection feedforward control tries to quickly realize a suitable injection timing based only on the measured intake air mass flow, intake manifold pressure, and engine speed. The A/F feedback control compensates the unavoidable errors in the feedforward loop. The boundedness of these parameter estimates depends on the convergence of the A/F steady-state error $\tilde{\lambda}$.

In the following, according to the Lyapunov stability theory for control systems, the stability and convergence analysis will be given for the closed-loop system with the adaptive fuel injection control. First, a positive-definite function is defined as

$$V = \frac{1}{2}(\lambda - \lambda_s)^2 \tag{4.34}$$

Then, its time derivative using the fuel model can be calculated as follows:

$$
\begin{aligned}
\dot{V} &= (\lambda - \lambda_s)\left(\frac{\ddot{m}_o}{\dot{m}_{fc}} - \frac{\dot{m}_o}{\dot{m}_{fc}^2}\ddot{m}_{fc}\right) \\
&= \tilde{\lambda}\frac{\dot{m}_o}{\dot{m}_{fc}^2}\left(\frac{1}{\lambda}\ddot{m}_o - (1-\varepsilon_w)\dot{u}_f - \tau_w u_f + \tau_w\frac{\dot{m}_o}{\lambda}\right)
\end{aligned}
\tag{4.35}
$$

Considering the dynamics of air mass flow and speed gives

$$
\begin{aligned}
\ddot{m}_o &= c_m \dot{p}_m \omega + c_m p_m \dot{\omega} \\
&= c_m\left[\frac{RT_m}{V_m}(\dot{m}_{th} - c_m p_m \omega)\omega + \frac{p_m}{J}(c_\tau p_m - D\omega - \tau_l)\right]
\end{aligned}
\tag{4.36}
$$

Moreover, since there is the uncertainty c_m associated with the volumetric efficiency η in $\dot{m}_o/\dot{m}_{fc}^2$, accordingly,

$$\frac{\dot{m}_o}{\dot{m}_{fc}^2} = \lambda^2 \frac{1}{\dot{m}_o} = \frac{\lambda^2}{p_m \omega}\frac{1}{c_m} \tag{4.37}$$

Substituting (4.36) and (4.37) into (4.35) yields

$$
\begin{aligned}
\dot{V} &= \tilde{\lambda}\frac{\lambda^2}{p_m \omega}\left[-\frac{1-\varepsilon_w}{c_m}\dot{u}_f - \frac{\tau_w}{c_m}u_f + \tau_w\frac{p_m \omega}{\lambda} + \frac{RT_m}{V_m}\frac{\omega}{\lambda}(\dot{m}_{th} - c_m p_m \omega)\right. \\
&\quad \left. + \frac{p_m}{J\lambda}(c_\tau p_m - D\omega - \tau_l)\right] \\
&= \tilde{\lambda}\frac{\lambda^2}{p_m \omega}\left[-\theta_1 \dot{u}_f - \theta_2 u_f + \theta_3\frac{p_m \omega}{\lambda} + \frac{RT_m}{V_m}\frac{\omega}{\lambda}(\dot{m}_{th} - \theta_4 p_m \omega)\right. \\
&\quad \left. + \frac{p_m}{J\lambda}(\theta_5 p_m - D\omega - \tau_l)\right]
\end{aligned}
$$

Choosing the fuel injection controller as (4.32) renders the time derivative of V to satisfy

$$\dot{V} = \tilde{\lambda} \frac{\lambda^2}{p_m \omega} \left[-\tilde{\theta}_1 \dot{u}_f - \tilde{\theta}_2 u_f + \tilde{\theta}_3 \frac{p_m \omega}{\lambda_s} - \theta_3 \frac{p_m \omega}{\lambda \lambda_s} \tilde{\lambda} - \tilde{\theta}_4 \frac{RT_m}{V_m} \frac{1}{\lambda} p_m \omega^2 \right. \\ \left. + \tilde{\theta}_5 \frac{1}{J\lambda} p_m^2 - k\tilde{\lambda} \right] \tag{4.38}$$

where $\tilde{\theta}_i = \theta_i - \hat{\theta}_i$, $(i = 1, 2, \cdots, 5)$.

Furthermore, a Lyapunov candidate function is chosen as

$$W = V + \frac{1}{2} \tilde{\theta}^T \Gamma^{-1} \tilde{\theta} \tag{4.39}$$

where $\tilde{\theta} = [\tilde{\theta}_1 \ \tilde{\theta}_2 \ \tilde{\theta}_3 \ \tilde{\theta}_4 \ \tilde{\theta}_5]^T$, $\Gamma = \text{diag}\{\gamma_i\}$ $(i = 1, 2, \cdots, 5)$; then, by considering (4.38) and the adaptive update laws (4.33), it is straightforward to show that the time derivative of W satisfies

$$\dot{W} = -k \frac{\lambda^2}{p_m \omega} \tilde{\lambda}^2 - \theta_3 \frac{\lambda}{\lambda_s} \tilde{\lambda}^2 \tag{4.40}$$

Thus, from (4.39) and (4.40), it can be concluded that $\tilde{\lambda}$ and $\tilde{\theta}$ are bounded by the Lyapunov stability theorem and $\tilde{\lambda} \to 0$ as $t \to \infty$ by the LaSalle invariant principle. Consequently, it follows that the proposed controller guarantees the convergence of $\tilde{\lambda}$ and the boundedness of the adaptive estimation parameters $\hat{\theta}$.

For the simplicity of the designed controller and its implementation in ECU, here the simplification of the fuel injection controller (4.32) will be discussed.

It can be noted that the fuel compensation term with the oxygen sensor feedback (the second and third terms) in the controller (4.32) is in fact equivalent to $\dot{m}_o/(c_m \lambda)$, which represents the amount of fueling expected to be lost or added for the corresponding variation in the airflow rate out of the manifold. Meanwhile, according to the relation between the air mass flow rate into the manifold $\dot{m}_{th}$ and the air mass flow rate out of the manifold $\dot{m}_o$, described as (4.19), it also can be noted that the variation of $\dot{m}_o$ is equivalent to the variation of $\dot{m}_{th}$ at the static operation mode, and at the transient mode the difference between the variation of $\dot{m}_o$ and the variation of $\dot{m}_{th}$ lies in the variation of p_m. Accordingly, the variation of $\dot{m}_o$ will be approximated by the variation of $\dot{m}_{th}$. Furthermore, the stoichiometric value λ_s and the estimate $\hat{\theta}_4$ will be used instead of λ and c_m as a result of the feedback transportation time delay of the oxygen sensor λ and the uncertainty of parameter c_m. Consequently, the adaptive fuel injection controller (4.32) is simplified as

$$\hat{\theta}_1 \dot{u}_f + \hat{\theta}_2 u_f = \hat{\theta}_3 \frac{p_m \omega}{\lambda_s} + \frac{\dot{m}_{th}}{\hat{\theta}_4 \lambda_s} + k\tilde{\lambda} \tag{4.41}$$

where the adaptive update laws on estimates $\hat{\theta}_1, \hat{\theta}_2, \hat{\theta}_3, \hat{\theta}_4$ are the same as those in (4.33).

Remark 4.6   The simplified controller (4.41) retains two elements in the original controller (4.32): the steady-state fueling term for the essential amount of required fuel estimated by the measured engine speed and manifold pressure, as well as the estimate of the uncertain parameter (the first term), and the feedback regulation term for absolute steady-state A/F accuracy (the last term). The substitute term (the second term) is a simple feedforward form, which means the estimation for the variation amount of fueling expected to be lost or added converted by the variation rate of the air charge into the manifold, instead of the variation rate of the air charge into the cylinder, and in which not utilizing directly the measured information of the A/F λ provides roughly a remedy for the transport delay of λ.

Remark 4.7   It can be noted from (4.41) that the simplified controller is constructed only by the measurable engine speed, manifold pressure, air mass flow rate into the manifold, and A/F, as well as the adaptive parameter estimates designed. Furthermore, $\ddot{m}_{th}$ in the controller (4.41) can be implemented by $\ddot{m}_{th} = \frac{1}{T_s}(\dot{m}_{th}(kT_s) - \dot{m}_{th}((k-1)T_s))$, where T_s is the sample period of the algorithm operation in microcomputers, kT_s represents the kth sample time, and $(k-1)T_s$ the $(k-1)$th sample time. Consequently, the simplified controller (4.41) is a feasible scheme for the implementation on ECU.

4.5   Experimental Validation for Control Strategies

Validation experiments for all presented control schemes in this chapter are conducted on an engine test bench described in Section 1.3.

4.5.1   Case Studies for Air Charge Estimation

In this subsection, experimental case studies validate the effectiveness of the air charge estimation for the A/F control performance during the transient operation of engines. Namely, the emphasis of the test is on the dynamics of the intake manifold filling and emptying resulting from the adjustment of the throttle opening, the regulation of the transient A/F by using the estimated cylinder air charge and the air charge estimation-based control law.

The following two operation modes with the variations of the air mass flow rate passing through the throttle valve $\dot{m}_{th}$ and the intake manifold pressure p_m will be used to show the air charge estimation, the fuel injected mass of each cylinder, and the A/F.

Mode I (load variation): The variation of engine load torque and the corresponding variations of the air mass flow rate and the intake manifold pressure are shown in Figure 4.10a.

Mode II (acceleration–deceleration operation): The variation of engine speed and the corresponding variations of the air mass flow rate and the intake manifold pressure are shown in Figure 4.10b.

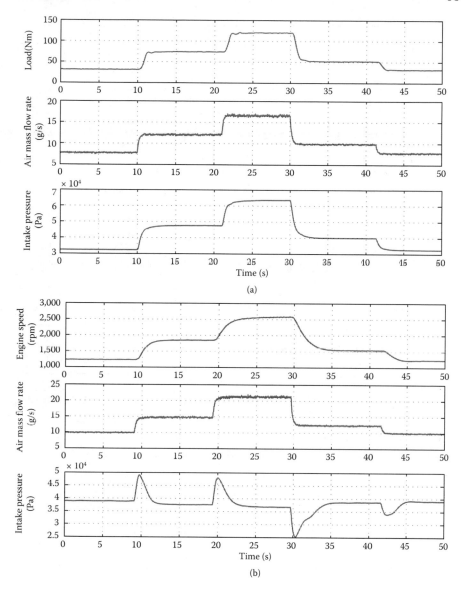

FIGURE 4.10
Operating modes for validating air charge estimation. (a) Mode I; (b) Mode II.

And then, the feedforward control law (4.7) and the air charge prediction-based control law (4.12) are implemented in both Mode I and Mode II, respectively.

Figure 4.11 shows the air charge estimation and the fuel injected mass of each cylinder and the left bank and right bank A/F for the feedforward

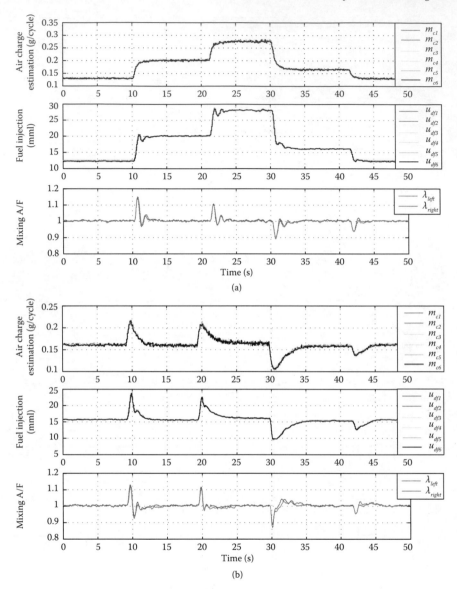

FIGURE 4.11
Results of air charge estimation and feedforward control. (a) Result of Mode I;
(b) result of Mode II.

control law (4.7); it should be noted that the A/F shown in the figure is
the normalized A/F. From the A/F response curves shown in Figure 4.11,
we can observe that for the feedforward control law the maximum positive
excursion of A/F is about 15.1% under the load variation from 30 to 75 Nm
and about 12.2% in acceleration from 1,200 to 1,800 rpm, and the minimum

negative excursion of A/F is about -11.4% under the load variation from 125 to 50 Nm and about -12.8% in deceleration from 2,600 to 1,500 rpm, respectively.

Figure 4.12 is for the prediction control law (4.12). Meanwhile, the above evaluated values for the prediction control law are about 12.1%, 10.2%, -9.6%, and -11.3%, respectively, as shown in Figure 4.12. The experimental validation results indicate that using the prediction controller, the A/F performance is relatively improved under the same situations.

4.5.2 Case Studies for Wall-Wetting Compensation

In this subsection, the testing experiments aim to validate the proposed compensation control design with respect to the switched dual-injection modes for the influences of engine speed and load on the fuel path dynamics.

Thus, in the test experiments, the engine operation conditions are chosen as engine speeds 1,600, 2,000, and 2,500 rpm and loads 80, 140, and 200 Nm, which can be achieved by operating the throttle and the dynamometer.

All the experiments are carried out under warmed-up conditions to the engine. In order to investigate the transient A/F response, the source injection controller, including the essential transient fueling compensation from ECU, is cutoff. The ignition timing is delivered by the source controller from the engine ECU. Moreover, experimental observation shows that at steady-state operation modes, the variations of engine speed and the airflow are minor when the authority of the injection command is changed, as shown in Figure 4.7; hence, the influence of air charge on the A/F is omitted. The injection commands delivered to the engine ECU are the individual direct injection and individual port injection, respectively, in milli-milliliters (mml) per cycle. The commands u_f^* for obtaining λ_s with respect to different engine speeds and loads are measured at steady operation modes.

First, the experimental identification for the model parameters τ_w and ε_w is given. The data sets are collected each sampled per 0.001 s, that is, with a sampling rate $T_s = 0.001$ s, around the switching time t_{Sk}. The index of port injection authority g_s is changed from 0 to 1 or from 1 to 0; in other words, the command is definitely switched between direct injection and port injection.

The model for identification is established as follows. Taking the wall-wetting delay, the intake-to-exhaust delay, and the sensor delay into account, the transfer function from a trigger impulse input signal δ at the switching time when the injection path changes to the response of the UEGO sensor at the gas mixing point can be represented by

$$G_\lambda(s) = \frac{y}{\delta} = \frac{\beta}{\tau s + 1} \cdot e^{-T_0 s} \cdot \frac{1}{\tau_0 s + 1} \qquad (4.42)$$

where $y = \lambda_s/\lambda - 1$, $\beta = \tau y(0)$, T_0 denotes the intake-to-exhaust delay, and τ_0 denotes the time constant of the sensor. For simplicity, consider that T_0 is determined by the engine speed ω, that is, $T_0 = 2\pi/\omega$ and $\tau_0 = 0.2$. The point

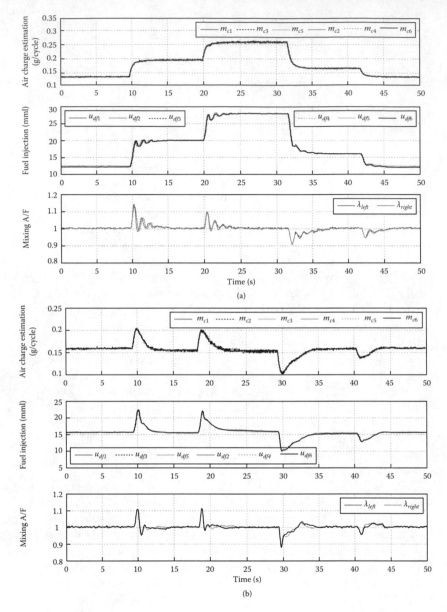

FIGURE 4.12

Results of air charge estimation and prediction control. (a) Result of Mode I;
(b) result of Mode II.

in each collected data set with the maximum absolute value is considered to be $y(0)$, and denote the corresponding timing as $t = t_0$.

In the time domain, the measured sensor response satisfies the following differential equation:

$$0.2\tau\ddot{y}(t) + (0.2 + \tau)\dot{y}(t) + y(t) = 0, \ t > t_0$$

Then, the discrete-time model to be identified is as follows:

$$y(t_{k+2}) - y(t_{k+1}) + \left(\frac{T_s}{0.2} - 1\right)(y(t_{k+1}) - y(t_k))$$

$$= -\frac{T_s}{\tau}\left(\frac{T_s}{0.2}y(t_k) + y(t_{k+1}) - y(t_k)\right) \tag{4.43}$$

where $t_k = kT_s \ (k=1, \cdots, n)$ denotes the sampling index. Using the sampled λ, the parameter τ of the model (4.43) is obtained with the least-square algorithm.

Finally, the values of τ_w and ε_w are calculated by $\tau_w = 1/\tau$, and $\varepsilon_w = 1 - 1/(1 + y(0))$ as direct injection is switched to port injection and, conversely, $\varepsilon_w = 1/(1 + y(0)) - 1$.

Figure 4.13 shows the measured A/F and the simulated model output of an identification experiment at engine speed 2,500 rpm and load 200 Nm, which indicates a good agreement between the measured λ and the simulated one.

The identification results are shown in Figure 4.14 for the various operation conditions.

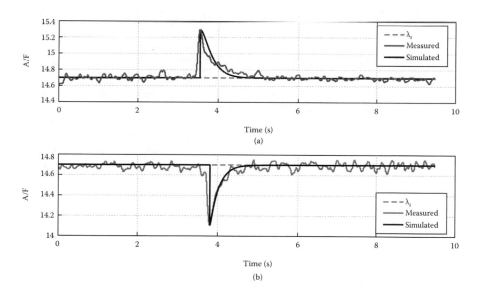

FIGURE 4.13
Comparison between model output and measured A/F: (a) direct injection switches to port injection and (b) port injection switches to direct injection.

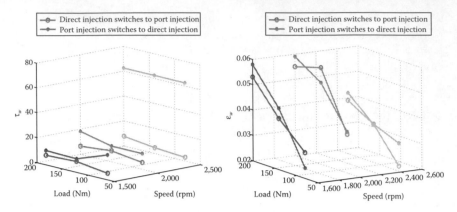

FIGURE 4.14
Identified τ_w and ε_w at different engine speeds, different loads.

From these results, we get the following points:

- Under the same operation conditions, the values of τ_w and ε_w for reducing the authority of port injection are bigger than for the inverse case.

- For the cases of increasing and reducing g_s, ε_w, and τ_w almost increase as the load is increased. It can be found that τ_w shows a few points (at 1,600 rpm/80 Nm with reducing g_s, and at 1,600 rpm/140 Nm and 2,000 rpm/140 Nm with increasing g_s) that do not comply with this rule. Moreover, it should be noted that if the engine is operated at higher speed, for example, 2,500 rpm, the load effect to τ_w is not obvious when the authority of the port injection is reduced. And one can note that the value of the wall-wetting time constant $1/\tau_w$ ($1/62.5 = 0.016$ s) is less than the engine cycle time 0.048 s.

- As speed increases, τ_w does not increase strictly, especially in the case of increasing the authority of the port injection, and ε_w does not strictly drop for the two cases of increasing and reducing port injection authority.

- It is worth pointing out that the influence of load to the parameter ε_w actually indicates that the variation of u_f^* at the switching time of injection commands is the main parameter affecting ε_w.

It is well known that the model (4.13) gives an average characterization for the port injection path dynamics. Using the identified model parameters shown in Figure 4.14, the generated injection compensation by (4.16) is delivered to each cylinder cyclically.

Experimental observation shows that using the identified τ_w and ε_w cannot achieve the desired compensation during the transient stage of changing

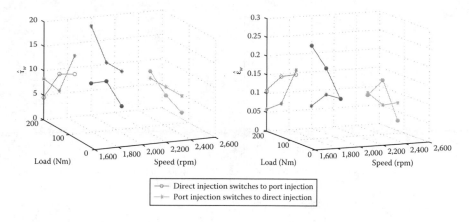

FIGURE 4.15

Calibrated τ_w and ε_w at different engine speeds, different loads.

the authority of the injection commands. By a trial-and-error method, the identified parameter values are calibrated to get improved compensation. Figure 4.15 shows the calibrated $\hat{\tau}_w$ and $\hat{\varepsilon}_w$, which are used in the validation experiments actually.

It can be observed that significant modifications to the parameters are actually given to ε_w when the engine operates at lower speed and with lighter load, and to τ_w when the engine operates at higher speed. Comparing with the identified parameter values, more bigger ε_w and more smaller τ_w are acceptable to achieve better transient compensations.

Corresponding experiments are conducted with switching port injection to direct injection definitely, and with 50% coordination between u_{fd} and u_{fp} to validate the dynamic compensator. First, a result at engine speed 1,600 rpm and load 90 Nm when port injection is definitely switched to direct injection is as shown in Figure 4.16. The validation results are shown in Figure 4.17 with a fixed load of 200 Nm and different engine speeds and in Figure 4.18 with a fixed engine speed of 1,600 rpm and different loads, respectively. t_1 through t_4 denote four mode switching times t_{Sk}. At t_1 and t_2, g_s is changed from 0 to 0.5 and to 0 again, respectively, and the compensation law α is not employed. Then, at t_3 and t_4, with the same change acting on g_s, the compensation law α is now working in the direct injection path. These results indicate the effectiveness of the compensation during the transient stage.

4.5.3 Case Studies for Adaptive Regulation

In this subsection, the testing experiments focus on the validation for the proposed adaptive controller consisting of the feedback/feedforward compensation and feedback regulation with the adaptive update laws.

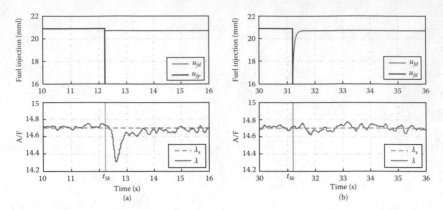

FIGURE 4.16

Case of port injection definitely switched to direct injection. (a) Without compensation; (b) with compensation.

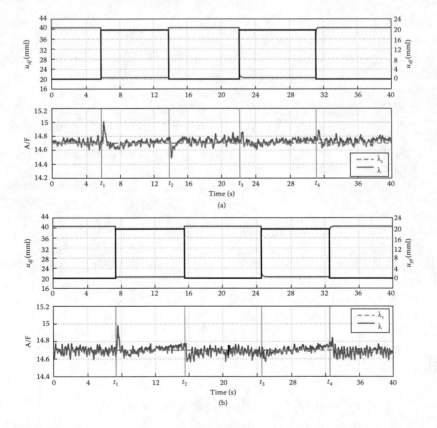

FIGURE 4.17

Validation results with 200 Nm load at different speeds. (a) At 2,000 rpm; (b) at 2,500 rpm.

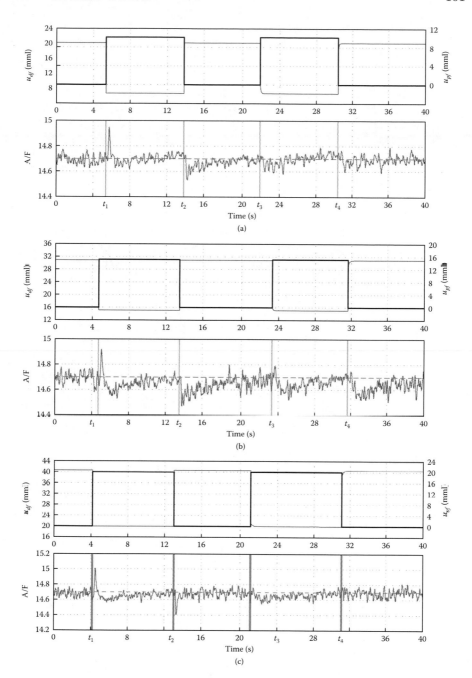

FIGURE 4.18

Validation results with different loads at 1,600 rpm. (a) With load 80 Nm; (b) with load 140 Nm; (c) with load 200 Nm.

   To illustrate the performance of the proposed controller with respect to the engine system enduring parameter variations under different conditions and load disturbances, as well as an acceleration–deceleration process, the following three cases are tested in experiments:

Case I: To validate the robustness of the adaptive controller to the parameter uncertainties compared with the fixed gain controller, the following experiment is conducted under the condition throttle angle $\phi = 9°$ and $\tau_l = 90$ Nm:

1. Before the timing t_1, the controller with the adaptive update laws operates properly.

2. At timing t_1, switch off the adaptive update laws (4.33) and reset $\hat{\theta}_i$ ($i = 1, 2, \cdots, 5$) in the controller to other constant values until the timing t_2. The constant values are the steady values recorded beforehand of the adaptive parameters at another operation mode ($\phi = 6°$ and $\tau_l = 60$ Nm), which means that in this period the fixed-gain controller operates and the system parameters vary.

3. At timing t_2, switch on the adaptive update laws.

Case II: To validate the effectiveness of the control scheme for the engine enduring load variations, the experiments are conducted as follows:

1. The engine system absorbs ±25% load variations from the dynamometer during 5 s; that is, the load torque τ_l is varied according to the shape shown in Figure 4.19a.

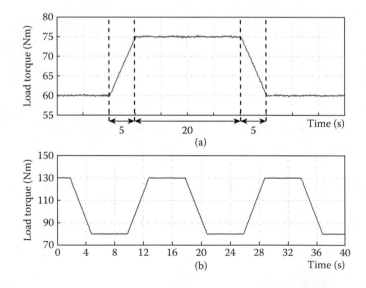

FIGURE 4.19

Load variations. (a) ±25% load variations during 5 seconds. (b) Large and steep load variations from 130 Nm to 80 Nm during 3 seconds.

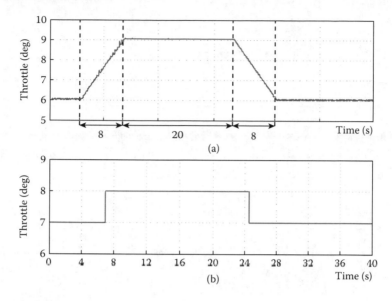

FIGURE 4.20
Throttle variations. (a) Slow the acceleration-deceleration operation. (b) Fast acceleration-deceleration operation.

2. The engine endures large and steep load variations, that is, load from 130 to 80 Nm during 3 s and back to 130 Nm during 3 s, as shown in Figure 4.19b.

Case III: To validate the effectiveness of the control scheme during the acceleration–deceleration operation of the engine:

1. Slow the acceleration–deceleration operation: A ramp with 3° variations in 8 s of throttle opening command is given, as shown in Figure 4.20a.

2. Fast acceleration–deceleration operation: A step-type and small-amplitude throttle opening command is given, as shown in Figure 4.20b.

ECU accepts fuel injection command U_f in milli-milliliter (mml) per cycle for each cylinder, that is, the discrete command to the actuator in practice. Hence, for implementation, the injection command delivering to ECU is calculated according to the following equation with the obtained fuel mass flow rate u_f:

$$U_f(t_k) = \frac{1}{6} \cdot u_f(t_k) \cdot (t_k - t_{k-1})/\rho_f \tag{4.44}$$

where ρ_f denotes the density of the used gasoline, k denotes the cycle-based sampling index, and t_k denotes the time at the kth sampling timing. Indeed, Equation 4.44 proposes a method realized in the cycle-based mean-value sense.

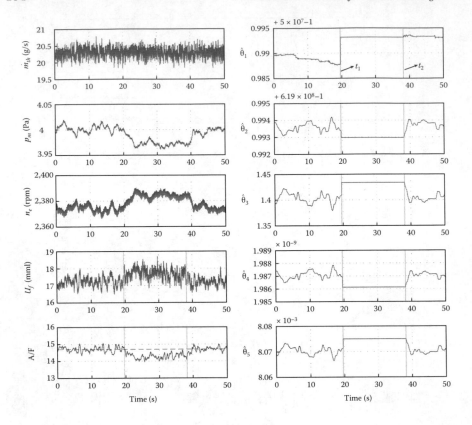

FIGURE 4.21
Experimental result with original controller of case I.

The desired A/F for each experiment is set as $\lambda_s = 14.7$, and the controller parameters are selected as follows:

$$\gamma_1 = 1 \times 10^6, \ \gamma_2 = 1.2 \times 10^5, \ \gamma_3 = 1 \times 10^{-2},$$
$$\gamma_4 = 1 \times 10^{-22}, \ \gamma_5 = 1 \times 10^{-8}, \ k = 5 \times 10^4$$

First, the tests of cases I, II-1, and III-1 are conducted on the engine test bench with the adaptive controller and the simplified one, respectively.

The inputs and outputs of the engine system, as well as the output of the adaptive update law, are shown in Figures 4.21 through 4.26, which include the response curves of the air mass flow $\dot{m}_{th}$, the manifold pressure p_m, the engine speed n_e, the injection command U_f, the A/F λ, and $\hat{\theta}_i$ $(i = 1, 2, \cdots, 5)$.

Results in Figures 4.21 and 4.22 show that both the original adaptive controller and the simplified one can guarantee the system operates well at the desired λ_s despite perturbations of the system parameters.

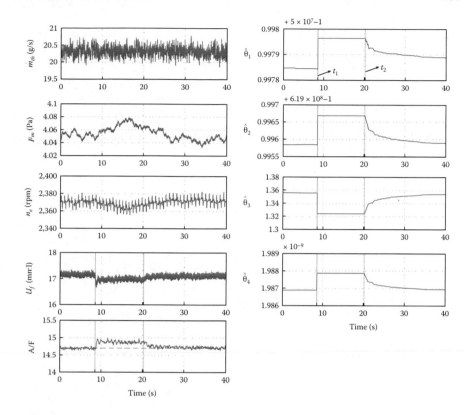

FIGURE 4.22
Experimental result with simplified controller of case I.

Concretely, the A/F response curves shown in the two figures before the timing t_1 and after the timing t_2 mean that the control system with adaptive laws can converge quickly to the desired A/F value λ_s, while the results in the two figures during $[t_1, t_2]$ show that there exists an A/F excursion for the controller without the adaptive update laws when there are variations of system parameters.

Results in Figures 4.23 and 4.24 show that A/F is convergent with a $\pm 5.8\%$ error boundary under the adaptive controller (4.32) and with a $\pm 1.02\%$ error boundary under the simplified controller (4.41) during the transient stage, respectively.

Results in Figures 4.25 and 4.26 show that the A/F can converge to the desired value with the performances $|\tilde{\lambda}| \leq 7.2\%\lambda_s$ and $|\tilde{\lambda}| \leq 1.35\%\lambda_s$ during the transient stage by the adaptive controller (4.32) and the simplified controller (4.41), respectively.

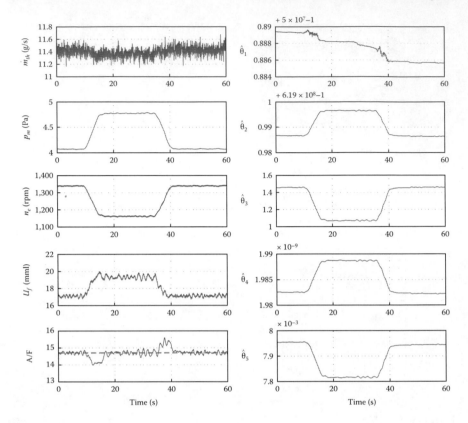

FIGURE 4.23
Experimental result with original controller of case II-1.

Moreover, the above experimental validation results indicate that using the simplified controller, the A/F performance is relatively improved under the same situations.

Thus, for the tests of cases II-2 and III-2, the control performance of the engine test bench with the simplified controller is only validated. The results are shown in Figures 4.27 and 4.28, respectively.

The result in Figure 4.27 shows that the fluctuation of the response to large and steep load variation is larger than that to small and slow load, compared with the result in case II-1, shown in Figure 4.24. The result in Figure 4.28 shows that the excursion of the A/F is at a transient stage for the step-type throttle opening (fast acceleration–deceleration), compared to the ramp-type throttle opening (slow acceleration–deceleration) in case III-1, shown in Figure 4.26.

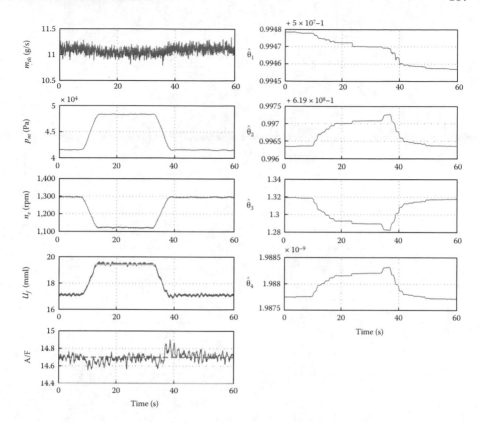

FIGURE 4.24
Experimental result with simplified controller of case II-1.

4.6 Conclusions

In this chapter, we mainly discussed the A/F control problem for gasoline engines according to their physical characteristics, particularly at transient operating. As is well known, the control input variable of the A/F control system is the fuel injection command, which is highly dependent on the cylinder air charge mass, the fuel injected dynamics decided by the injection modes, and the physical parameter uncertainties resulting from the extensive operation range of the engine. Thus, we presented the A/F control schemes targeting the three aspects, respectively.

First, the main target was to improve the transient performance of the A/F with the simple algorithm of cylinder air charge estimation. As a consequence, an estimation for cylinder air charge was developed that mainly focuses on the dynamics of the air path and utilizes a discrete-time TDC-scaled average form.

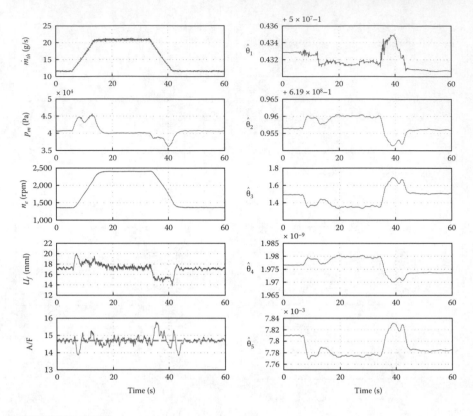

FIGURE 4.25
Experimental result with original controller of case III-1.

Meanwhile, a control scheme with multistep prediction was proposed to handle the delay from the fuel injection command to the fueling implementation, and a real-time adjusting law was introduced to compensate the modeling error and uncertainty. Unfortunately, it follows from the experimental results that the A/F excursion is rather large at transients. The sources probably include that the description for the exhaust and the sensor was only considered a first-order lag instead of its dynamics being modeled by identification or the observer technique, the delay d_2 from the fueling implementation to the beginning of the exhaust stroke was not considered in the prediction control, and an improved control algorithm should be incorporated into the fuel injection command. These remedies are expected to further research to improve the available results.

And then, for the gasoline engines with a dual-injection system, this work targets the injection control problem during the transient stage when the authority between the two injection commands is changed. On the basis of the mean-value model for characterizing the dynamics of the port injection

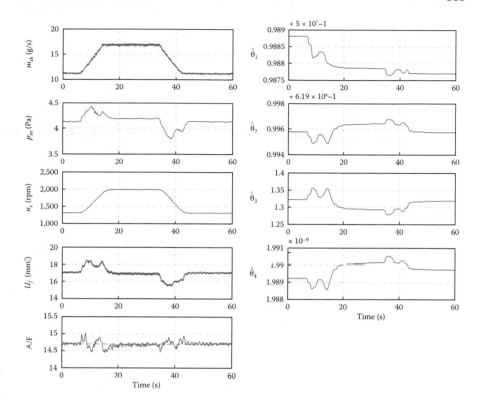

FIGURE 4.26
Experimental result with simplified controller of case III-1.

path, a dynamic feedforward compensation was proposed to adjust the direct injection to obtain the desired A/F during the transient stage of injection path changing. A distinct identification method was employed to obtain the parameters of the wall-wetting model. Experimental results demonstrated the efficiency of the compensation injection control approach.

Lastly, a mean-value model-based adaptive A/F control scheme was presented for gasoline engines that targets the parameter uncertainties in the model. The derivation of the proposed control law, which consists of the feedforward of air mass flow passing through the throttle valve and the feedback of the intake manifold pressure and the A/F sensor, was performed by following the physics of transient dynamics, including the intake manifold, the fuel injection path, and the exhaust-to-sensor output. The proof of convergence of the A/F regulation and the boundedness of parameter estimation are obtained under a Lyapunov-based stability analysis framework. Moreover, for the sake of simplicity in implementation, the proposed control scheme was simplified based on the observation of engine physics. To demonstrate the effectiveness of the presented control schemes, experiments conducted on a V6 gasoline

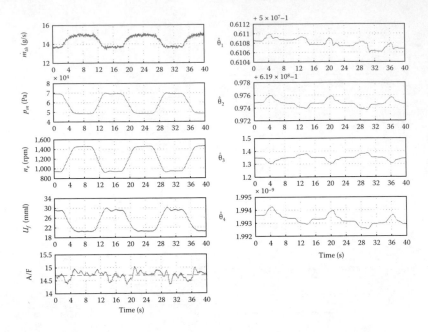

FIGURE 4.27
Experimental result with simplified controller of case II-2.

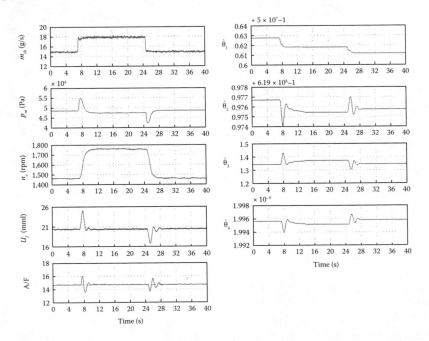

FIGURE 4.28
Experimental result with simplified controller of case III-2.

engine were demonstrated at several transient operating modes involving parameter uncertainties, load changes, acceleration–deceleration, and so forth. It should be noted that as a common drawback of adaptive control, the regulation performance is not satisfied sufficiently when the load or the demand of acceleration–deceleration is changed dramatically. This fact can be observed from the experimental results shown in Figure 4.28. Improving the control scheme to adapt quick changes of operating conditions is a further challenge.

5

Receding Horizon Optimal Control

5.1 Introduction

Receding horizon control (RHC), also known as model predictive control (MPC), has attracted wide attention in the automotive industry over the last decade. Relative to other conventional control approaches, RHC possesses the following significant advantages: (1) it is able to tackle the explicitly constrained optimization problem; (2) it can easily deal with the control problem with the time-delay system; (3) it is effective for solving multivariable and multiobjective optimal control problems; (4) the control scheme can be easily understood owing to its model-based design concept; and (5) it has good robustness under situations of disturbance, even with changing system parameters.

The fundamental principle of RHC is to use the future state of the control plant predicted based on the dynamical model to optimize the current control input, while the optimization process is achieved by solving the optimal control problem iteratively over a finite time horizon with the given cost function. In essence, there are three central concepts for RHC: the model-based prediction, the finite time horizon optimization, and the continuous moving horizon strategy. For instance, at each control timing that the control input is decided, the system state is measured and taken as the initial condition of the predicted model, and then a sequence of the optimal control input is obtained by solving an open-loop optimization problem over the given prediction horizon to minimize the error between the predictive output and reference command. However, only the first value of the optimal control sequence is implemented to the control plant until the next control timing starts. As a result, the system state is detected in real time and the same optimization process is repeated with the above receding horizon concept. Such a feedback optimization procedure is shown in Figure 5.1.

In fact, RHC theory has achieved much success in industrial fields since the first practical application was exploited in the petrochemical industry in the early 1970s. The early implementation was mainly based on a simple step-response model for predicting the future process dynamics, and it was typically named dynamic matrix control (DMC). However, DMC has poor performance with handling the interactive multivariable control plants [62]. Thereafter, more applications with RHC concepts have emerged,

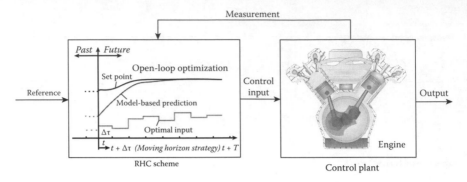

FIGURE 5.1
Basic concept of receding horizon optimal control.

such as model algorithmic control (MAC), extended horizon adaptive control (EHAC), and generalized predictive control (GPC) [63]. It should be noted that these control designs usually derive the explicit analytical solution owing to the linear system structure. However, if the control plants have strictly nonlinear characteristics, the control problem with receding horizon optimization concepts will become very complicated because it leads to solving the Hamilton–Jacobi–Bellman (HJB) partial differential equation. Therefore, industrial application with RHC to the nonlinear system is always a challenging research topic.

Nevertheless, some active explorations for handling such nonlinear RHC application problems have been conducted recently, and two kinds of approaches are usually adopted by researchers. One is to convert the nonlinear system model to a linear one with some proper assumptions, or by using the feedback linearization technique. Indeed, in some control cases, the system dynamics can be approximately treated with linear property in certain and narrow operating ranges, that is, engine idling speed control [20]. Instead of the optimal solution, another approach uses the approximate numerical solution by means of methods such as particle swarm optimization (PSO) [64], the sequential quadratic programming (SQP) method [65], and the sensitivity function iteration method [66]. The common bottlenecks for these numerical algorithms, including the complicated programmability and considerable computing load, actually restrict their wide application in the industrial field. Recently, a numerical optimization algorithm named the continuation and generalized minimum residual estimation (continuation/GMRES) method [67] has attracted more attention owing to its fast computing capacity and easy implementation. This optimization method can solve the linear equation instead of the HJB differential equation in real time by means of a combination of the continuation solving algorithm and the generalized minimum residual method.

Back to the topic of the engine transient control, engine system is indeed a physical plant with the presence of limitations on itself and the actuators. It is also a sophisticated control system involving many control loops to achieve single or multiple objectives; these features are motivating the development of RHC scheme in practical engine control issues.

In this chapter, we investigate several nonlinear RHC-based application cases, including the engine torque control and engine speed control problems. First, a unified RHC design framework for these engine control problems is introduced based on the continuation/GMRES on-line optimization algorithm. The continuation/GMRES algorithm can provide an approximately optimal solution for the nonlinear RHC controllers in real time. Then, two individual RHC controllers aimed to achieve the accurate torque tracking and speed tracking control are designed, respectively. The transient control performance in wide engine operating range is the main focus. Moreover, a parameter tuning approach for the RHC scheme is proposed, and based on it, we give the further application for speed tracking control.

In this study, the control-oriented model is derived from the simplified mean-value model of the gasoline engine. Specifically, the nonlinear predictive model is directly applied to the torque tracking control, and for speed tracking control, the on-line RHC algorithm is applied with nonlinear feedback compensation. The experimental studies for torque tracking control and speed tracking control are implemented on a full-scale gasoline. The transient performance and robustness of the control system are evaluated on the engine-in-the-loop simulation system for torque control experimental studies, and the idling speed control and wide range speed tracking control are conducted for verifying the speed tracking controller.

5.2 Design Framework and Optimization Algorithm

As previously mentioned, RHC can be essentially classified into the scope of optimal control. Consider a dynamical system represented by

$$\dot{x}(t) = f(x(t), u(t)) \tag{5.1}$$

where $x \in \mathbb{R}^n$ and $u \in \mathbb{R}^m$ denote the state and control inputs, respectively; a performance index for the receding horizon optimal control problem is typically formulated as follows:

$$J(u) = \Phi\big(x(t+T)\big) + \int_t^{t+T} L\big[x(\tau), x_d(\tau), u(\tau)\big]d\tau \tag{5.2}$$

where $0 \leq \tau \leq T$, $T > 0$, is the predictive horizon length and $x_d(\tau)$ is the desired state trajectory.

Suppose that the system state $x(t)$ at time t can be obtained. The receding horizon control problem is to find a function $u^*(t+\tau)$ such that the performance index (5.2) takes the minimum along the trajectory of system (5.1) forced by the function $u^*(t+\tau)$ under the given constraints

$$C(x^*(t+\tau), u^*(t+\tau)) = 0 \tag{5.3}$$

where $C(\cdot)$ is a p-dimensional continuous vector function, and at time t only the first value of the optimal control trajectory is taken as the final control input, that is,

$$u(t) = u^*(t), \quad \tau = 0$$

The block diagram of the RHC scheme is illustrated in Figure 5.2. In fact, at each sampling time, the instant system state will be fed back to the controller, and it is taken as the new initial values of the optimization problem. At the same time, the optimal control input will be applied to the control plant. The optimization process can be used to solve the finite horizon optimal control problem with a free end point in each sampling period and repeatedly taking place in the moving time domain.

For the sake of simplicity, we denote $x_t(\tau) = x(t+\tau)$ and $u_t(\tau) = u(t+\tau)$; then the finite horizon optimization problem can be described in the following. For the given initial state $x(t)$ at time t, the optimal control input trajectory $u_t^*(\tau)$ is obtained by minimizing the given performance index (5.2), that is,

$$u_t^*(\tau) = \arg\min\{J(u)\} \tag{5.4}$$

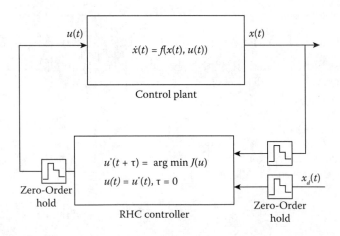

FIGURE 5.2

The feedback control structure of RHC.

subject to the system dynamics, initial conditions, and constraints:

$$\begin{cases} \dot{x}_t^*(\tau) = f(x_t^*(\tau), u_t^*(\tau)) \\ x_t^*(0) = x(t) \\ C(x_t^*(\tau), u_t^*(\tau)) = 0 \end{cases} \tag{5.5}$$

We now discuss how to solve this optimal control problem. Due to the presence of the path constraints on the control input or state, the proposed finite horizon optimization problem can be reduced to the two-point boundary-value problem (TPBVP) by adding the constraint to the performance index J as [70]

$$\bar{J}(u) = \Phi\big(x(t+T)\big) + \int_t^{t+T} H\big[x(\tau), u(\tau), \lambda(\tau), \mu(\tau)\big] d\tau \tag{5.6}$$

where $H(\cdot)$ is the Hamiltonian defined by

$$H(x, \lambda, u, \mu) = L(x, x_d, u) + \lambda^{\mathrm{T}} f(x, u) + \mu^{\mathrm{T}} C(x, u)$$

and $\lambda^{\mathrm{T}} \in \mathbb{R}^n$ and $\mu^{\mathrm{T}} \in \mathbb{R}^p$ denote the co-state vector and Lagrange multiplier related to the equality constraint, respectively. Then the problem is finally reduced to identify a stationary point of $\bar{J}$, and the optimal solution must satisfy the following necessary conditions:

$$\dot{x}_t^*(\tau) = f\big(x_t^*(\tau), u_t^*(\tau)\big) \tag{5.7}$$

$$x_t^*(0) = x(t) \tag{5.8}$$

$$\dot{\lambda}_t^*(\tau) = -H_x^{\mathrm{T}}\big(x_t^*(\tau), x_d^*(\tau), \tilde{u}_t^*(\tau), \lambda_t^*(\tau), \mu_t^*(\tau)\big) \tag{5.9}$$

$$\lambda_t^*(T) = \Phi_x^{\mathrm{T}}\big(x_t^*(T)\big) \tag{5.10}$$

$$H_u^{\mathrm{T}}\big(x_t^*(\tau), x_d^*(\tau), u_t^*(\tau), \lambda_t^*(\tau), \mu_t^*(\tau)\big) = 0 \tag{5.11}$$

$$C\big(x_t^*(\tau), u_t^*(\tau)\big) = 0 \tag{5.12}$$

Considering the discrete-time property of the programming algorithm in practice, the above receding horizon optimal problem is always solved with the discrete-time form. Basically, the predictive horizon length T can be divided into N steps; then the discrete form of the dynamic system can be easily deduced from the continuous-time model (5.7) by means of the forward difference method, which is shown as follows:

$$x_t^*(k+1) = x_t^*(k) + f(x_t^*(k), u_t^*(k))\Delta\tau \tag{5.13}$$

where

$$x_t^*(k) = x^*(t + k\Delta\tau), \ u_t^*(k) = u^*(t + k\Delta\tau), \quad k = 0, 1, \cdots, N$$

and $\Delta\tau = T/N$ is the sampling interval.

Moreover, the necessary conditions (5.9) through (5.12) can be discretized as

$$\lambda_t^*(k) = \lambda_t^*(k+1) + H_x^{\mathrm{T}}\big(x_t^*(k), x_d^*(k), u_t^*(k), \lambda_t^*(k+1), \mu_t^*(k)\big)\Delta\tau \qquad (5.14)$$

$$\lambda_t^*(N) = \Phi_x^{\mathrm{T}}\big(x_t^*(N)\big) \qquad (5.15)$$

$$H_u^{\mathrm{T}}\big(x_t^*(k), x_d^*(k), u_t^*(k), \lambda_t^*(k+1), \mu_t^*(k)\big) = 0 \qquad (5.16)$$

$$C\big(x_t^*(k), u_t^*(k)\big) = 0 \qquad (5.17)$$

It should be noted that the approximately optimal control sequence $[u_t^*(0), u_t^*(1), \ldots, u_t^*(N-1)]$ can be obtained by solving the above discrete equations. And this optimal solution will converge to the solution of the original continuous-time optimization problem as $N \to \infty$ under mild conditions [62].

Then, collect all the control inputs and multipliers over the predictive horizon steps into one vector as follows:

$$U_t = [u_t^{*\mathrm{T}}(0), \mu_t^{*\mathrm{T}}(0), \cdots, u_t^{*\mathrm{T}}(N-1), \mu_t^{*\mathrm{T}}(N-1)]^{\mathrm{T}}$$

In fact, for a given U_t and initial condition $x_t(0)$, system states $x_t^*(k)$, $(k = 0, \ldots, N)$ can be calculated recursively by Equation 5.13, and then $\lambda_t^*(k)$ can be reversely calculated from N-step to 0-step by Equations 5.14 and 5.15. Herein, the determination of U_t becomes the critical point, but unfortunately, U_t is difficult to obtain since it must satisfy the following equality conditions derived from (5.16) and (5.17):

$$F(U_t, x_t, t) := \begin{bmatrix} H_u^{\mathrm{T}}(x_t^*(0), x_d^*(0), u_t^*(0), \lambda_t^*(1), \mu_t^*(0)) \\ C(x_t^*(0), u_t^*(0)) \\ \vdots \\ H_u^{\mathrm{T}}(x_t^*(N-1), x_d^*(N-1), u_t^*(N-1), \lambda_t^*(N), \mu_t^*(N-1)) \\ C(x_t^*(N-1), u_t^*(N-1)) \end{bmatrix}$$

$$= 0 \qquad (5.18)$$

That is, U_t is coupling with the values of $\{x_t^*(k)\}_{k=0}^N$ and $\{\lambda_t^*(k)\}_{k=0}^N$. An awkward approach for solving Equation 5.18 is to recursively calculate $\{x_t^*(k)\}_{k=0}^N$ and $\{\lambda_t^*(k)\}_{k=0}^N$ with the assumed U_t by cut and trial. However, such a process requires powerful computational capacity, and it is unfeasible in the real-time control system.

To improve the computing efficiency with few iteration times, a continuation calculation concept can be applied by only putting the focus on the time derivative of U such that (5.18) is satisfied. According to continuation method, instead of solving equation $F(U_t, x_t, t) = 0$ itself at each time, we just choose the proper initial value U_0 and take the time derivative of Equation 5.18 in top account. Specifically, the computational process can be represented by

$$F(U_0, x_t(0), 0) = 0$$
$$\dot{F}(U_t, x_t, t) = -\zeta F(U_t, x_t, t)$$

where $\zeta > 0$. Furthermore, the above equations can be written as follows if F_{U_t} is nonsingular:

$$\dot{U}_t = F_{U_t}^{-1}(-\zeta F - F_{x_t}\dot{x}_t - F_t) \tag{5.19}$$

where the Jacobians F_{U_t}, F_{x_t}, and F_t are obtained by forward difference approximations and the GMRES method in view of the computational load. As an iterative method for numerical solution of the nonsymmetric system of linear equations, GMRES can approximate the solution by the vector in a Krylov subspace with minimal residual. It can obtain the approximate solution with fewer iterative calculations [68, 69]. Combining the continuation method and GMRES, $\dot{U}_t$ can be obtained and U_t is accordingly calculated by integrating $\dot{U}_t$ in real time. In practical programs, suppose the sampling period is Δt, and then the whole algorithm can be concluded as follows:

General Work Flow of Continuation/GMRES Method

Step 1: Initialize time $t := 0$; measure the state $x_0 := x_t(0)$ at current initial time t; find an initial $U_0 = U_t(0)$ analytically or numerically such that $||F(U_0, x_0, 0)|| \le \delta$ for the positive δ.

Step 2: Extract control sequence u^{T} from U_t; for $t' \in [t, t + \Delta t)$, set $u(t') = u^{\mathrm{T}}$.

Step 3: At the next sampling time $t + \Delta t$, measure the state $x_{t+\Delta t}(0)$. Set $\Delta x = x_{t+\Delta t}(0) - x_t(0)$.

Step 4: Based on $\Delta x, U_t, \widehat{U}_t$, and $x_t(0)$, compute $\dot{U}_t$ using the GMRES methods; here the initial guess $\widehat{U}_t$ is chosen referring to the following forms: $\widehat{U}_t = 0$ or $\widehat{U}_t = \widehat{U}_{t-\Delta t}$ with $\widehat{U}_{t-2\Delta t} = 0$.

Step 5: Set $U_{t+\Delta t} = U_t + \dot{U}_t \Delta t$.

Step 6: Set $t = t + \Delta t$.

Step 7: Go back to step 2.

Besides, the optimization problems with inequality constraints are more significant in practice. In most cases, the inequality constraints $|u(\tau)| \le u_{\mathrm{lim}}$

can be equivalently converted to equality constraints by introducing a dummy input $u'(\tau)$ as follows [70]:

$$C_1(u(\tau), u'(\tau)) = u^2(\tau) + u'^2(\tau) - u_{\lim}^2 = 0 \qquad (5.20)$$

It means that no matter what value $u'(\tau)$ takes, the inequality $|u(\tau)| \le u_{\lim}$ must be satisfied. u_t and u'_t are independent of each other in the whole optimization process. Thus, the new control input vector at time t becomes $\tilde{u}_t = [u_t, u'_t]^T$.

Since u'_t cannot be updated when $u'_t = 0$, a small dummy penalty term $-ru'$ $(r > 0)$ is added in the Hamiltonian as follows:

$$H = L(x, x_d, u) - ru' + \lambda^T f(x, u) + \mu^T C(x, u) + \mu_1^T C_1(\tilde{u}) \qquad (5.21)$$

In addition, the form of the additional term $-ru'(\tau)$ is not arbitrary. For example, if $-ru'(\tau)$ is modified as $ru'^2(\tau)$, then the larger r is, the more restricted $u'(\tau)$ gets. Finally, to make the performance index take the minimum value, $u'(\tau)$ has to equal zero, that is, $u(\tau) = u_{\lim}$. There is no meaning in doing this. What is more, the performance index decides the values of the control input. In (5.21), r is a small positive constant, which can hardly affect the optimality. That is the purpose with a small constant. This is easily proved by analyzing the optimality condition (5.11). The optimal solution with inequality conditions also follows the aforementioned optimization algorithm; therefore, we omit explicit mention of it.

It should be noted that the proposed optimization algorithm provides a general design framework for the receding horizon control, and also, an approximate optimal solution can be derived in each control period. Based on this design framework, the engine torque controller and speed controller will be introduced.

5.3   Torque Transient Control

Engine torque is one of the most important control variables in modern engine control systems, and the torque demand control strategy is becoming a consensus in view of the increasing complexity of the control logic and the interacting influences on the different engine subsystems [82, 83]. In such a control strategy, a torque demand decision block is needed to manage the torque requirements from the driver and the other assistant control units; then it gives the synthetic torque demand value to the torque tracking controller. A schematic of the torque demand strategy is shown in Figure 5.3. Based on the current operating conditions and the reference torque, the torque tracking controller will adjust the throttle opening angle to achieve the desired torque. Obviously, the accurate and fast control performance for engine torque tracking is crucial to realize the torque demand strategy. In what follows, the RHC-based torque tracking control will be discussed.

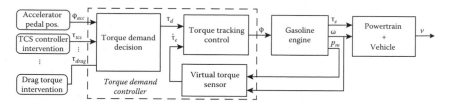

FIGURE 5.3
A schematic of the torque demand strategy.

5.3.1 Control-Oriented Model

Although RHC strongly depends on the model quality, an accurate engine dynamical model is actually complicated to derive because it involves thermodynamics, fluid mechanics, and mechanical kinematics. In practical control design, analysis of engine dynamics is usually dealt with in the sense of mean value and control orientation in order to reduce the order of the control system and avoid unnecessary complexity. For a receding horizon optimization algorithm, the real-time computing load will increase with the complexity of the model. In this regard, the control-oriented models that focus on the intake air charging dynamics and the crankshaft rotational dynamics are provided by means of the mean-value modeling approach, as well as the curving fitting technique.

As mentioned in Section 2.5, the dynamics of the intake manifold pressure is easily deduced according to the ideal gas equation in the following form:

$$\dot{p}_m = \frac{RT_m}{V_m}(\dot{m}_{th} - \dot{m}_o) \qquad (5.22)$$

To avoid the modeling complexity, the air mass flow rate $\dot{m}_{th}$ can be written approximately in the following expression, if holding the assumption that the air fluid passing through the throttle orifice is incompressible:

$$\dot{m}_{th} = c_d \cdot A_{th}(\phi) \cdot \sqrt{2\rho_a} \cdot \sqrt{p_a - p_m} \qquad (5.23)$$

where c_d is the discharge coefficient. Besides, $A_{th}(\phi)$ is the opening area of the throttle valve following the relationship (2.16).

Since the air mass flow rate going into the cylinder can be given based on the mean value model (2.51), the intake air charging dynamics can be finally represented by

$$\dot{p}_m = a_0(1 - \cos\phi)\sqrt{p_a - p_m} - a_2\omega p_m \qquad (5.24)$$

where $a_0 = \frac{RT_m}{V_m}\pi r^2 c_d\sqrt{2\rho_a}$. For the sake of simplicity, a_0 and a_2 can be approximately regarded as constant.

Engine torque production is a complicated physical process influenced by many factors, such as fuel injection amount, spark timing, and air–fuel ratio.

However, for simplification purposes, we assume the engine is well controlled by different control loops and ignore the intake-to-power delay, and then the engine torque can be represented as a function with respect to the engine speed and intake manifold pressure in a mean-value sense [81], that is,

$$\tau_e = g_1(\omega)p_m + g_2(\omega) \tag{5.25}$$

where $g_1(\omega)$ and $g_2(\omega)$ are the variable parameters corresponding to the engine speed, and they can be calibrated from the engine static tests.

In addition, the rotational dynamics of the crankshaft can be easily deduced according to Newton's law, and it is formulated as

$$\dot{\omega} = \frac{1}{J}(\tau_e - \tau_l) \tag{5.26}$$

It should be pointed out that the rotational dynamics can be ignored in a certain predictive horizon for the torque controller design in consideration that the time constant of rotational dynamics is much larger than that in the dynamics process of torque production. In fact, the speed is regarded as a constant during each predictive horizon. Hence, the control-oriented model for torque tracking control is mainly determined by (5.24) and (5.25).

5.3.2   Torque Tracking Control Scheme Design

Let us now focus on the torque tracking controller design with these control-oriented models. Note that the purpose of the torque tracking control is to find an optimal control input for the throttle angle such that the engine generated torque can track any given desired torque command with sufficient response speed. Based on the receding horizon optimization algorithm, a candidate of the performance index for the torque tracking problem is chosen as follows:

$$J(\dot{\phi}) = \int_t^{t+T_\tau} [r_{\tau 1}(\tau_d(\tau) - \tau_e(\tau))^2 + r_{\tau 2}\dot{\phi}(\tau)^2]d\tau \tag{5.27}$$

where $\tau \in [0, T_\tau]$, T_τ is the predictive horizon, τ_d denotes the desired torque command, and $r_{\tau 1}, r_{\tau 2}$ are the weighting coefficients for the tracking error and control inputs, respectively. In addition, the control input $\dot{\phi}$ actually represents the change rate of the throttle angle ϕ, and it is introduced here instead of penalizing the throttle angle itself for avoiding over-sensitive action of the throttle angle.

Based on the above cost function (5.27), the receding horizon torque optimal control law is given at each time t $(t \geq 0)$,

$$\dot{\phi}(t) = \dot{\phi}_t^*(0)$$

where $\dot{\phi}_t^*(\tau)$ is defined as $\dot{\phi}^*(t+\tau)$, and it is obtained by minimizing the cost function (5.27) subject to the intake air charging dynamics (5.24), torque production model (5.25), and following physics constraints (5.28):

$$\begin{cases} p_{m\,\min} \leq p_m(t) \leq p_{m\,\max} \\ \phi_{\min} \leq \phi(t) \leq \phi_{\max} \\ u_{\min} \leq \dot{\phi}(t) \leq u_{\max} \end{cases} \tag{5.28}$$

The above discussion presents a typical receding horizon optimization problem formulation. Moreover, to on-line solve such an optimization problem with the continuation/GMRES method, a slight change for the control input is applied, and the system dynamics is extended to a two-dimensional system.

Denote u as an auxiliary control signal and let $u = \dot{\phi}$; the throttle angle ϕ is therefore determined by the integral of u once it is provided by the optimization solution. Equivalently, the modified control law in the fashion of the receding horizon optimal control is decided by

$$u(t) = u_t^*(0) \tag{5.29}$$

with $u_t^*(\tau)$ that minimizes the following performance index,

$$J(u_t) = \int_t^{t+T_\tau} [r_{\tau 1}(\tau_d(\tau) - \tau_e(\tau))^2 + r_{\tau 2} u_t^2(\tau)] d\tau \tag{5.30}$$

subject to the system dynamics (5.31),

$$\begin{cases} \dot{p}_m = a_0(1 - \cos\phi)\sqrt{p_a - p_m} - a_2 \omega p_m \\ \dot{\phi} = u \end{cases} \tag{5.31}$$

and the proposed inequality constraints (5.28).

It should been noted that the notation $u_t(\tau)$ is used for $u(t+\tau)$ defined on $\tau \in [0, T_\tau]$ for the sake of simplicity. Similarly, the symbols with the subscript t in what follows are defined as a form similar to that of $u_t(\tau)$, and we will omit any explicit mention of it.

In the real-time implementation, the system states $p_m(t)$ and $\dot{\phi}(t)$ at time t are fed back to the optimization algorithm, and then the optimal $u_t^*(\tau)$ at time t is provided in the discrete-time sequence with N steps by exploiting the continuation/GMRES method, that is,

$$\{u_t^*(0), u_t^*(T_0), \cdots, u_t^*((N-1)T_0)\}$$

where T_0 is the RHC optimization discrete-time intervals yielded to $T_0 = T_\tau/N$. The first value of the above control sequence is finally applied to the engine control system. The controller structure is illustrated in Figure 5.4.

Embedded Integrator. Note that an important factor for influencing the above controller performance is the precision of the predictive model; that is, it mainly depends on the accuracy of the nonlinear intake air charging

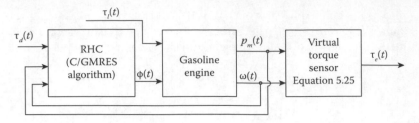

FIGURE 5.4
Torque tracking controller structure.

dynamic model. However, it is well known that an accurate model is difficult to be obtained, especially for the nonlinear engine system. In order to further improve the torque tracking accuracy, an integrator of the tracking error is embedded in the control loop. In fact, it is well known that embedding an integrator is an effective way to compensate the modeling error and external disturbance.

In the content below, we further apply the RHC scheme to the integrator-embedded plant with the same problem formulation in the following procedure.

First, define the throttle angle in the following form:

$$\phi(t) = k_\tau e_I(t) + v(t) \tag{5.32}$$

where $e_I(t) = \int_0^t (\tau_d(\tau) - \tau_e(\tau)) d\tau$ denotes the integrator of the torque tracking error, $v(t)$ is introduced as an auxiliary control input, and k_τ is the integral gain.

Then, the system dynamics with the embedded integrator can be represented as

$$\begin{cases} \dot{p}_m = a_0 \left(1 - \cos(k_\tau e_I + v)\right) \sqrt{p_a - p_m} - a_2 \omega p_m \\ \dot{e}_I = \tau_d - \tau_e \end{cases} \tag{5.33}$$

Repeating a similar argument, the modified control law for the torque tracking problem can be obtained by

$$v(t) = v_t^*(0) \tag{5.34}$$

with $v_t^*(\tau)$ deduced by solving the following finite horizon optimization problem,

$$\min_{v_t^*(\tau)} \left\{ \int_t^{t+T_\tau} [r_{\tau 1}(\tau_d(\tau) - \tau_e(\tau))^2 + r_{\tau 2} v_t^2(\tau)] d\tau \right\} \tag{5.35}$$

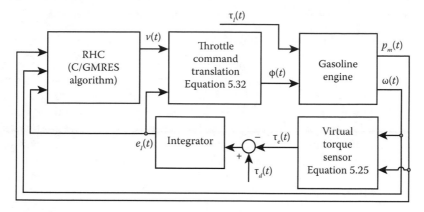

FIGURE 5.5
Torque tracking controller with embedded integrator.

subject to the dynamical system model (5.33) and the following constraints:

$$\begin{cases} p_{m\,\min} \le p_m(t) \le p_{m\,\max} \\ \phi_{\min} \le \ \phi(t) \ \le \phi_{\max} \\ v_{\min} \le \ v(t) \ \le v_{\max} \end{cases} \tag{5.36}$$

In this design framework, the integrator of the tracking error e_I is mainly used for eliminating the steady error, while the additional control variable v is able to improve the transient tracking performance. Finally, the optimal throttle angle can be obtained according to (5.32) once the optimal auxiliary control input $v(t)$ is derived by (5.34). The controller structure is illustrated in Figure 5.5.

It should be pointed out that the proposed torque control method is based on the single control variable design. In fact, an alternative design method with multiple control variables is easily derived where the integral gain can be also treated as another control variable to be optimized in real time. Under such an extension design, the control input of the optimization problem becomes a vector with two elements:

$$\mathbf{u}_t(\tau) = \begin{bmatrix} v_t(\tau) \\ k_{\tau t}(\tau) \end{bmatrix}$$

Accordingly, the performance index is modified as

$$J(\mathbf{u}_t(\tau)) = \int_t^{t+T_\tau} [r_{\tau 1}(\tau_d(\tau) - \tau_e(\tau))^2 + r_{\tau 2}v_t^2(\tau) + r_{\tau 3}k_\tau^2(\tau)]d\tau \tag{5.37}$$

This multiple variable design method can be applied with the same on-line optimization algorithm; the detailed problem formulation is not repeated here.

5.4   Speed Transient Control

Engine speed control is the fundamental control problem and remains important because it has great influence on the fuel economy and emission
performance. Indeed, idle speed control has aroused wide attention in past
research. Generally, the engine speed in idling conditions mainly keeps in a
low and narrow operating range, where the nonlinearity is not very apparent in
comparison with the wide operating range. Therefore, many linear controllers
have obtained better control performance for tackling such control problems.

Recently, with the development of automotive technology, more and more
studies have begun to focus on the speed tracking problem in a wide engine
operating range. Especially, such wide-range speed tracking control is significant for the coordinative control of hybrid powertrains [71] and gearshift
control of the automatic mechanical transmission [72, 73]. In this context, this
section will design a speed controller that is not only for the idling regulation
but also for the wide-range tracking control.

5.4.1   Tracking Error Dynamics

From the observation of physics, the dynamical response from the throttle to
speed variation is mainly determined by the intake air charging dynamics and
the crankshaft rotational dynamics. As discussed in Section 5.3, the physical
model for speed tracking controller design can be concluded as follows:

$$\begin{cases} \dot{p}_m = a_0(1 - \cos\phi)\sqrt{p_a - p_m} - a_2\omega p_m \\ \dot{\omega} = \dfrac{1}{J}\big(g_1(\omega)p_m + g_2(\omega) - \tau_l\big) \end{cases} \tag{5.38}$$

The speed tracking control is aimed to find the optimal throttle opening $\phi(t)$ such that the actual engine speed $\omega(t)$ tracks any given reference
speed $\omega_d(t)$ even in a wide operating range, with accurate and fast tracking
performance. Refer to the design method of the torque tracking controller;
a similar receding horizon optimization framework can be proposed based on
the above nonlinear system dynamics (5.38). However, such a design is tedious
to be mentioned here again. Alternatively, we adopt another design method
for speed tracking control.

To achieve speed tracking for any given command, we would like to convert
the physical dynamical model to a tracking error dynamical system. For the
given desired speed command ω_d and actual speed ω at time t_0, a reference
speed model is designed in the following form to provide a smooth reference
speed trajectory $\omega_r(t)$:

$$\omega_r(t) = \delta^{(t-t_0)}\omega(t_0) + (1 - \delta^{(t-t_0)})\omega_d(t_0) \tag{5.39}$$

where $t \geq t_0$, $\delta = \exp(-1/T_f)$, and T_f is the time constant of the reference
trajectory.

With the reference speed trajectory, the speed tracking error is defined as

$$\begin{cases} e_1 = \omega - \omega_r \\ e_2 = \dot{\omega} - \dot{\omega}_r \end{cases} \tag{5.40}$$

Then the tracking error dynamics can be obtained by combining (5.38) and (5.40):

$$\begin{cases} \dot{e}_1 = e_2 \\ \dot{e}_2 = \alpha_0(\omega, p_m, \phi) + \beta_0(\omega, p_m)\tau_l \end{cases} \tag{5.41}$$

where

$$\alpha_0(\omega, p_m, \phi) = -\beta_0(\omega, p_m)\big(g_1(\omega)p_m + g_2(\omega)\big) - \ddot{\omega}_r$$
$$+ \frac{1}{J} g_1(\omega)\big(a_0(1 - \cos\phi)\sqrt{p_a - p_m} - a_2\omega p_m\big)$$

$$\beta_0(\omega, p_m) = -\frac{1}{J^2}\left(\frac{\partial g_1(\omega)}{\partial\omega}p_m + \frac{\partial g_2(\omega)}{\partial\omega}\right)$$

Moreover, introduce an additional control variable u defined as

$$u = \alpha_0(\omega, p_m, \phi) + \beta_0(\omega, p_m)\tau_l \tag{5.42}$$

Then the Equation 5.41 becomes

$$\begin{cases} \dot{e}_1 = e_2 \\ \dot{e}_2 = u \end{cases} \tag{5.43}$$

It should be noted that for any given reference trajectory (5.39), the tracking error dynamics (5.41) is now represented by a second-order linear model (5.43) under nonlinear compensation (5.42). Accordingly, the final throttle angle ϕ is obtained:

$$\phi = \arccos(\gamma_1(\omega, p_m, u) + \gamma_2(\omega, p_m)\tau_l) \tag{5.44}$$

where

$$\gamma_1(\omega, p_m, u) = 1 - \frac{J}{a_0 g_1(\omega)\sqrt{p_a - p_m}}\left(u + \ddot{\omega}_r + \frac{a_2 g_1(\omega)}{J}\omega p_m - \frac{1}{J^2}(g_1(\omega))\frac{\partial g_2(\omega)}{\partial\omega}\right.$$
$$\left. - g_2(\omega)\frac{\partial g_1(\omega)}{\partial\omega})p_m - \frac{1}{J^2}g_1(\omega)\frac{\partial g_1(\omega)}{\partial\omega}p_m^2 - \frac{1}{J^2}\frac{\partial g_2(\omega)}{\partial\omega}g_2(\omega)\right)$$

$$\gamma_2(\omega, p_m) = \frac{J}{a_0 g_1(\omega)\sqrt{p_a - p_m}}\left(\frac{\partial g_1(\omega)}{\partial\omega}p_m + \frac{\partial g_2(\omega)}{\partial\omega}\right)$$

The above transformation is actually a state feedback linearization used in nonlinear control theory [74].

5.4.2   Speed Tracking Control Scheme Design

Now let us discuss the control design based on the derived tracking error dynamics. Similarly to the torque tracking control design, an integrator can be embedded into the predictive model in order to eliminate the steady-state tracking error caused by an insufficiently accurate model. In this regard, let us define the control input u consisting of two items:

$$u(t) = k_s e_3(t) + \vartheta(t) \qquad (5.45)$$

where $e_3(t) = \int_0^t e_1(\tau)d\tau$ and k_s is the integral gain of the speed tracking error; $\vartheta(t)$ is an auxiliary control input to be designed later.

Then the system dynamics with the embedded integrator consists of three states, which are represented as

$$\begin{cases} \dot{e}_1 = e_2 \\ \dot{e}_2 = k_s e_3 + \vartheta \\ \dot{e}_3 = e_1 \end{cases} \qquad (5.46)$$

Recall the framework of the receding horizon control design; the constraint of $\vartheta(t)$ is taken into account and complies with

$$\vartheta_{\min} \leq \vartheta(t) \leq \vartheta_{\max} \qquad (5.47)$$

Then the receding horizon optimization problem can be formulated as follows. For the given reference speed command $\omega_d(t)$, the optimal control action $\vartheta(t)$ for speed tracking control at each time t can be obtained by

$$\vartheta(t) = \vartheta_t^*(0) \qquad (5.48)$$

with $\vartheta_t^*(\tau)$ that minimizes the following performance index and is subject to the dynamic system (5.46) and constraint (5.47).

$$J(\vartheta_t) = r_{s1}e_1^2(t+T_s) + \int_t^{t+T_s} [r_{s2}e_{1t}^2(\tau) + r_{s3}\vartheta_t^2(\tau)]d\tau \qquad (5.49)$$

where T_s is the predictive horizon; r_{s1}, r_{s2}, and r_{s3} are the weighting coefficients. The optimal throttle opening can be derived from (5.45) once the optimal $\vartheta(t)$ is provided. The expression of the throttle angle is described by

$$\phi(t) = \arccos\left[1 - \frac{J\Theta(\vartheta, p_m, \omega)}{a_0 g_1(\omega(t))\sqrt{p_a - p_m(t)}}\right] \qquad (5.50)$$

where

$$\Theta(\vartheta, p_m, \omega) = \vartheta(t) + k_s e_3(t) + \ddot{\omega}_r(t) + \frac{1}{J}a_2\omega(t)p_m(t)$$
$$- \frac{1}{J^2}\left(\frac{\partial g_1(\omega)}{\partial \omega}p_m(t) + \frac{\partial g_2(\omega)}{\partial \omega}\right)$$
$$\times \left(g_1(\omega(t))p_m(t) + g_2(\omega(t)) - \tau_l(t)\right)$$

The general control structure for speed control is presented in Figure 5.6.

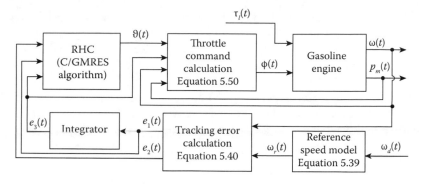

FIGURE 5.6
Block diagram of RHC-based speed control structure.

As before, the integral gain in this study can also be regarded as another control input to be optimized in real time. Then the cost function for this multiple-control-variable RHC design becomes

$$J(\mathbf{v}_t) = r_{s1}e_{1t}^2(T_s) + \int_t^{t+T_s} [r_{s2}e_{1t}^2(\tau) + r_{s3}\vartheta_t^2(\tau) + r_{s4}k_{st}^2(\tau)]d\tau \quad (5.51)$$

where $\mathbf{v}_t(\tau) = [k_s(\tau), \vartheta_t(\tau)]^{\mathrm{T}}$. Also, it can be applied with the same on-line optimization algorithm, so we omit mentioning it again.

5.5 Parameter Tuning

Receding horizon control provides a relatively simple approach to determine the feedback optimal control for linear or nonlinear systems, and the control performance is determined by not only the model precision, but also the controller parameters, such as weighting coefficients in the cost function. In fact, for a closed-loop control system, the weighting coefficients have a significant impact on the trade-off between the control response speed and control energy consumption. To adjust the control performance, the traditional parameter tuning approach usually adopts a cut-and-trial method by conducting many comparative tests. This approach is effective but time-consuming and redundant, especially for the high-dimension system. For this reason, a convenient parameter tuning approach has important engineering significance in consideration of time and cost.

Recently, a parameter tuning approach for receding horizon control performance was presented in [75] and [76], and indeed, its inspiration was obtained from the inverse linear quadratic (ILQ) regulator design. In general, this method is able to provide a single tuning parameter for the weighting coefficients while guaranteeing the optimality of the feedback control system. In this section, based on the given parameter tuning approach, we investigate the relations between the RHC control performance and the weighting parameters, and propose a practical parameter tuning application on the speed tracking problem.

Regarding the aforementioned controller design method, the control-oriented model used in the following speed tracking problem should be declared again only based on the feedback linearized model (5.43). Actually, Equation 5.43 represents second-order single-input linear system, and it can be further written in the following matrix form:

$$\dot{\mathbf{e}} = A\mathbf{e} + Bu \qquad (5.52)$$

where

$$A = \begin{bmatrix} 0 & 1 \\ 0 & 0 \end{bmatrix}, \quad B = \begin{bmatrix} 0 \\ 1 \end{bmatrix}, \quad \mathbf{e} = \begin{bmatrix} e_1 \\ e_2 \end{bmatrix}$$

According to the above linear system and referring to the aforementioned RHC design framework, the speed tracking control problem can be formulated as follows. For the given desired speed $\omega_d(t)$ at t time, the reference speed trajectory $\omega_r(t)$ is given by (5.39), and the optimization problem is to find an optimal control input $u^*(t)$ that is determined by

$$u^*(t) = u_t(0) \qquad (5.53)$$

where $u_t(\tau) = u(t + \tau)$, and it is obtained by minimizing the following performance index,

$$J(u) = \mathbf{e}^T(t+T)P\mathbf{e}(t+T) + \int_t^{t+T} [\mathbf{e}^T(\tau)Q\mathbf{e}(\tau) + Ru^2(\tau)]d\tau \qquad (5.54)$$

subject to the system dynamics (5.52) and the following constraint:

$$u_{\min} \le u(t) \le u_{\max} \qquad (5.55)$$

where $\tau \in [0, T]$, where T denotes the predictive horizon length, and $P \in \mathbb{R}^2 \ge 0, Q \in \mathbb{R}^2 \ge 0$, and $R \in \mathbb{R}^1 > 0$ are the quadratic weighting coefficients in the performance index. Therefore, the final control law for the throttle angle is determined by (5.44) when the optimal $u^*(t)$ is obtained.

For the above linear feedback optimal control system, let us now apply the given ILQ design-based parameter tuning approach to the selection of the quadratic weighting coefficients P, Q, and R.

Typically, the classical linear quadratic (LQ) regulator deals with the optimization problem in the infinite horizon domain with the following performance index,

$$J = \int_0^\infty (\mathbf{e}^T(\tau)Q\mathbf{e}(\tau) + Ru^2(\tau))d\tau \qquad (5.56)$$

and the final optimal cost holds with the solution of the Riccati equation P as

$$J = \mathbf{e}^T(0)P\mathbf{e}(0) \tag{5.57}$$

In fact, with the result given in (5.57), the performance index in the LQ problem can easily extend to that of RHC owing to the presence of the terminal cost, that is, the terminal cost in (5.54) is regarded as the infinite horizon cost in the interval $(t+T, \infty)$:

$$\mathbf{e}^T(t+T)P\mathbf{e}(t+T) = \int_{t+T}^{\infty} (\mathbf{e}^T(\tau)Q\mathbf{e}(\tau) + Ru^2(\tau))d\tau \tag{5.58}$$

Substituting (5.58) into (5.54), we can obtain the same expression as (5.56). That means, the optimization problem between the finite horizon and the infinite horizon can be exchangeable only if the weighting coefficient in terminal cost is the solution of the Riccati equation. Therefore, the tuning problem for RHC weightings can be essentially put into the framework of the ILQ design.

Now let us briefly review the inverse design of the linear quadratic regulator. For the given feedback control law $u = -K\mathbf{e}$, find the proper weightings Q and R such that the given feedback control matrix K is optimal and minimizes the performance index (5.56). In general, the ILQ design method does not need to specify the weighting coefficients for the controller design, while it will give some relations or conditions between quadratic weightings, the feedback control matrix, and system matrices to ensure the feedback law is optimal.

For instance, the proposed K is stable and optimal if and only if the follow conditions are satisfied for some $P > 0$ and $R > 0$ [77]:

$$\begin{cases} Q = H^TP + PH > 0 \\ RK = B^TP \end{cases} \tag{5.59}$$

where $H = \frac{1}{2}BK - A$ must be co-positive or diagonal dominant. Moreover, satisfactory P and R can be obtained by introducing some auxiliary matrices V and Σ and described as follows [78]:

$$P = (VK)^T D\Sigma^{-1}(VK) + Y \tag{5.60}$$

$$R = V^T DV \tag{5.61}$$

where V is the real nonsingular matrix and $\Sigma > 0$ is a real diagonal matrix. Actually, V and Σ depend on the real left eigenvalues and nonnegative eigenvalues of KB, respectively. $D > 0$ and $Y \geq 0$ must satisfy the conditions $\Sigma D = D\Sigma$ and $YB = 0$.

Note that the above conditions give the specific forms for determination of P, Q, and R. However, a challenging problem is how to choose the satisfactory parameters, such as V, D, Y, and K. To this end, a relatively simple method is proposed in [77], and we have summarized some main points in the following.

For an n-dimension control system with m control inputs, assume the feedback gain matrix K can be given in the following partition form:

$$K = \begin{bmatrix} \underbrace{K_1}_{n-m} & \underbrace{K_2}_{m} \end{bmatrix} \qquad (5.62)$$

Then the system matrices can be accordingly partitioned by

$$A = \begin{bmatrix} A_{11} & A_{12} \\ A_{21} & A_{22} \end{bmatrix}, \quad B = \begin{bmatrix} 0 \\ B_2 \end{bmatrix} \qquad (5.63)$$

where $\det(B_2) \neq 0$ and the dimensions of the matrices $A_{11}, A_{12}, A_{21}, A_{22}$, and B_2 are $n-m \times n-m$, $m \times n-m$, $n-m \times m$, $m \times m$, and m, respectively.

Based on the necessary conditions of optimality (5.59), K is optimal if and only if KB can be formulated as $KB = V^{-1}\Sigma V$ [77].

Let $K_1 = K_2 F_1$ and $K_2 = B_2^{-1} V^{-1} \Sigma V$. Then K can be parameterized as follows [79]:

$$K = B_2^{-1} V^{-1} \Sigma V [F_1 \quad I_m] \qquad (5.64)$$

where F_1 is the real matrix that relies on the partial pole placement. The determination algorithm for F_1 can be found in [79], which can guarantee the K is always optimal with the arbitrary nonsingular V and some appropriate Σ. Moreover, Σ can be further represented by

$$\Sigma = \sigma \Lambda, \quad (\sigma > 0) \qquad (5.65)$$

where $\Lambda > 0$ is a constant diagonal matrix.

Then the feedback gain K is mainly determined by the only positive parameter σ for the constant V. Accordingly, if providing the satisfactory D and Y, the weights P, Q, and R based on (5.59) through (5.61) are also only related to the single parameter σ.

Obviously, the proposed inverse design method provides a very convenient tuning parameter for the control performance of the closed-loop system, instead of the traditional cut-and-trial method on the weights P, Q, and R. Therefore, it can be applied to the tuning of the RHC performance.

Now recall the aforementioned speed tracking control problem, the general tuning procedure for RHC performance can be summarized as follows:

1. Partition the system matrices as the proposed form in (5.63). In this case, $A_{11} = 0$, $A_{12} = 1$, $A_{21} = 0$, $A_{22} = 0$, and $B_2 = 1$.

2. Specify $(n{-}m)$ stable poles that are different from the eigenvalues of A_{11}, and then design the matrix F_1 based on the algorithm in [79]. Then F_1 is obtained as 1.25 when the partial stable pole is chosen to be -1.

3. Choose the arbitrary nonsingular V and positive Λ, and then the lower bound of σ can be found by using the algorithm in [79]. Herein, we set $V = 1$ and $\Lambda = 1$, and the lower boundary of σ is calculated

to be 2.5149. For the sake of simplicity, the feedback matrix K is only related with the parameter σ, and it can be further written as follows:

$$K(\sigma) = [1.25\sigma \quad \sigma] \tag{5.66}$$

4. Choose the tuning parameter σ larger than its lower bound, and calculate the feedback gain K and the corresponding weights P, Q, and R. Since K is determined, the weightings P, Q, and R can be obtained by

$$P(\sigma) = (VK(\sigma))^T D\Sigma^{-1}(VK(\sigma)) + Y \tag{5.67}$$

$$Q(\sigma) = H^T P + PH \tag{5.68}$$

$$R = V^T DV \tag{5.69}$$

where $H = \frac{1}{2}BK(\sigma) - A$, and H must be a co-positive or diagonal dominant matrix; $D > 0$ and $Y \geq 0$ must satisfy the conditions $\Sigma D = D\Sigma$ and $YB = 0$. In this case, we choose

$$D = 0.1, \quad Y = \begin{bmatrix} 0.4 & 0 \\ 0 & 0 \end{bmatrix}$$

5. Solve the receding horizon optimal control problem by using the given weights.

6. Check the control performance of the closed-loop system and tune σ to obtain the desired response.

For clarity, the tuning procedure from the fourth step to the sixth step is illustrated in Figure 5.7.

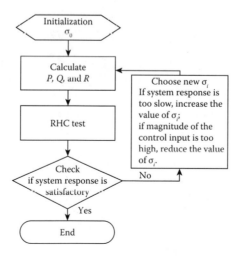

FIGURE 5.7
Tuning procedure for RHC performance.

5.6   Adaptive Compensation of Disturbance

In Section 5.5, the speed tracking controller design is based on the feedback
linearized model (5.43), and then the response speed of the closed-loop system
can be tuned by the single parameter σ. On the other hand, the control effects
depend on the model precision, as mentioned many times before. Also, there
is no integrator of tracking error embedded in the predictive model to compensate the control error. Thus, for the proposed speed controller design in
Section 5.5, some other methods must be designed to achieve accurate tracking
performance. To this end, an adaptive compensation algorithm is introduced
in this section.

Let us go back to the physical model of the engine speed (5.38). Suppose
the modeling error can be represented as an equivalent external disturbance
that acts on the mechanical rotational dynamics, that is,

$$\begin{cases} \dot{\omega} = \frac{1}{J}(g_1(\omega)p_m + g_2(\omega) - \tau_l - \epsilon) \\ \dot{p}_m = a_0(1 - \cos\phi)\sqrt{p_a - p_m} - a_2\omega p_m \end{cases} \tag{5.70}$$

where ϵ is the unknown constant disturbance.

In order to reject the disturbance, we construct an adaptive controller so
as to replace the unknown disturbance by an estimated parameter $\hat{\epsilon}$; then
under the linear feedback control law $u = -K\mathbf{e}$, the final throttle angle can
be derived from (5.44):

$$\begin{cases} \phi = \arccos[\gamma_1(\omega, p_m, -K\mathbf{e}) + \gamma_2(\omega, p_m)(\tau_l + \hat{\epsilon})] \\ \dot{\hat{\epsilon}} = \psi(\mathbf{e}, \hat{\epsilon}) \end{cases} \tag{5.71}$$

where $\psi(\mathbf{e}, \hat{\epsilon})$ is the adaptive law to be designed.

Let $\tilde{\epsilon} = \epsilon - \hat{\epsilon}$ and repeat the feedback linearization method argument in
Section 5.4.1. The closed-loop system with tracking error dynamics can be
represented in the following form by combining (5.70) and (5.71):

$$\begin{cases} \dot{\mathbf{e}} = (A - BK)\mathbf{e} - \beta_0(\omega, p_m)B\tilde{\epsilon} \\ \dot{\tilde{\epsilon}} = -\dot{\hat{\epsilon}} = -\psi(\mathbf{e}, \hat{\epsilon}) \end{cases} \tag{5.72}$$

Actually, it is easy to deduce the following results for this adaptive law,
which guarantee the closed-loop system to be stable in the Lyapunov sense.

Proposition 5.1   *For the closed-loop system (5.72), if the adaptive law is
given by*

$$\psi(\mathbf{e}, \hat{\epsilon}) = -\frac{1}{\rho}\beta_0(\omega, p_m)PB\mathbf{e} \tag{5.73}$$

with the adjusting gain $\rho > 0$, then the feedback system is asymptotically stable.

Proof. For the stability analysis of the system (5.72), the Lyapunov function can be chosen as follows:

$$U(\mathbf{e}, \tilde{\epsilon}) = \frac{1}{2}\mathbf{e}^T P \mathbf{e} + \frac{1}{2}\rho\tilde{\epsilon}^2 \qquad (5.74)$$

Then along the trajectory of the system (5.72), we have

$$\dot{U}(\mathbf{e}, \tilde{\epsilon}) = \frac{1}{2}\dot{\mathbf{e}}^T P \mathbf{e} + \frac{1}{2}\mathbf{e}^T P\dot{\mathbf{e}} + \rho\tilde{\epsilon}\dot{\tilde{\epsilon}}$$

$$= \frac{1}{2}\mathbf{e}^T[(A - BK)^T P + P(A - BK)]\mathbf{e} - \beta_0(\omega, p_m)PB\mathbf{e} - \rho\tilde{\epsilon}\dot{\tilde{\epsilon}}$$

$$= \frac{1}{2}\mathbf{e}^T[(A - BK)^T P + P(A - BK)]\mathbf{e} - (\beta_0(\omega, p_m)PB\mathbf{e} + \rho\psi(\mathbf{e}, \hat{\epsilon}))\tilde{\epsilon}$$

Note that from the Riccati equation, it can be derived that

$$(A - BK)^T P + P(A - BK) = \bar{Q} \qquad (5.75)$$

where $\bar{Q} = -(PBR^{-1}B^T P + Q) < 0$. Therefore, to guarantee the Lyapunov stability of the system, we can choose

$$\psi(\mathbf{e}, \hat{\epsilon}) = -\frac{1}{\rho}\beta_0(\omega, p_m)PB\mathbf{e} \qquad (5.76)$$

such that $\dot{U} \leq 0, \forall \mathbf{e}, \tilde{\epsilon}$. Furthermore, note that

$$\Omega_0 = \{(\mathbf{e}, \tilde{\epsilon}) \mid \dot{U} = 0\} \subseteq \Omega_e = \{\mathbf{e} = 0\}$$

Then under the LaSalle's invariant principle, we can conclude that the state $(\mathbf{e}, \tilde{\epsilon})$ converges to the maximum invariant set M included in the set Ω_0 as $t \to \infty$. This follows $\mathbf{e}(t) \to 0$ as $t \to \infty$.

Based on the above discussion, the control structure for the speed tracking control can be sketched in Figure 5.8.

Note that the presented adaptive control scheme focuses on the unknown disturbance that is caused by uncertainty in the load and modeling. The adaptive law (5.73) is designed based on the constant matrices P and B and the nonlinear time-varying function $\beta_0(\omega, p_m)$. In practice, the physical parameters included in $\beta_0(\omega, p_m)$, such as the inertia J and the torque generation model (5.25), are difficult to handle exactly. Especially, representing the torque generation by the relationship (5.25) is with strong uncertainty. It is recommended that nominal values of the physical parameters should be used for the adaptive law design. It should be pointed out that the time-varying uncertainty that cannot be equivalently represented by unknown constant disturbance will be out of handling by the presented adaptive control law.

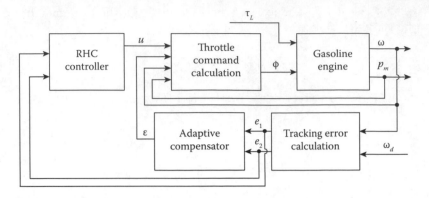

FIGURE 5.8
Block diagram of the speed tracking controller with adaptive compensator.

5.7   Experimental Case Studies

So far, we have proposed several receding horizon optimal controller designs, including torque tracking control, speed tracking control, and the parameter tuning approach for control performance. In fact, all the derived RHC controller schemes follow the unified design framework, and the on-line optimization process can be achieved by the proposed continuation/GMRES algorithm. In this section, we verify the proposed RHC controllers in the full-scaled gasoline engine; meanwhile, the transient control performance and real-time optimization capacity will be evaluated.

5.7.1   Torque Transient Control

For torque tracking control, we proposed two kinds of design schemes in the previous discussion, including a basic receding horizon optimal controller based on the proposed engine mean-value model and an improved controller with an embedded integrator for tracking error. Both controllers will be verified in the experiments to compare the control effects. The control objective is to adjust the throttle angle such that the engine-generated torque follows any given targeted command with sufficient response speed. To this end, a series of typical test commands, which consist of the step and ramp signals, is applied to verify the transient control performance. Meanwhile, to verify the robustness and effectiveness of the controllers in a wide operating range, we change the engine speed in real time by adjusting the dynamometer speed output. Thus, in the implementation, the engine speed actually follows a pseudosinusoidal change all the time. The experimental results with the typical commands are shown in Figure 5.9.

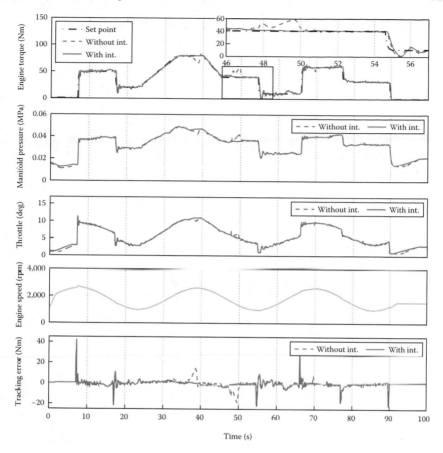

FIGURE 5.9
Experimental results for typical test commands by using RHC torque tracking controllers.

In the experiment, the predictive horizon lengths for both RHC controllers are chosen as 0.3 s in consideration that the process dynamics from throttle variation to torque generation are involved as possible. Besides, the control period is 0.01 s. That is, there are 30 steps that need to be calculated iteratively in one control period according to the discretized optimization solving method adopted in the continuation/GMRES algorithm. Moreover, as for the improved controller with embedded integrator of the tracking error, the integral gain k_{ι} is set as 0.3. The necessary constraints in the controllers are subject to the following configuration:

$$\begin{cases} 0.01 \leq p_m \leq 0.102 \ \ (\text{MPa}) \\ 0 \leq \phi \leq 86° \\ -200 \leq u \leq 200 \ (\text{without integrator}) \\ -3 \leq v \leq 3 \ \ (\text{with integrator}) \end{cases}$$

Figure 5.9 illustrates the comparative control effects of the proposed two torque tracking controllers. It can be also observed that, for both controllers, the actual engine torque can generally track the torque demand command quickly, even if the engine speed condition is varying in real time. The tracking error is within the margin of 40 Nm at most when the step command takes place. However, the control effects of the controller without integrator may exhibit a larger tracking error than those of the controller with an embedded integrator, especially in some operating conditions. One explanation for this difference can be reduced to the insufficiently accurate predictive model, as is proposed previously. Actually, the controller with integrator obtains better control performance, with the minor tracking error owing to an embedded integrator. That is, embedding an integrator is proven effective for tackling such model-based control with an imprecise model or external disturbance.

To further verify the robustness and effectiveness of the torque tracking controller with an embedded integrator, as well as the real-time performance of the on-line optimization algorithm, we conducted a driving cycle test with the engine-in-the-loop experimental configuration. The control structure of the engine-in-the-loop simulation system was introduced in Chapter 1. In this case, the virtual driver will drive the virtual vehicle to follow a given driving cycle speed with a real gasoline engine as a power source. The driver model uses a proportional-integral-differential (PID) control scheme, and the control structure of the system is sketched in Figure 5.10.

The experimental results with the engine-in-the-loop setup are shown in Figure 5.11. In the figure, the first subplot illustrates the vehicle speed following performance under the driver-demanded torque. Here, the demanded torque τ_d is given by a static function with respect to the driver accelerator pedal signal and engine speed, as shown in the second graph. The torque tracking controller calculates the proper throttle angle and controls the actual engine torque $\tau_e Actl$ to track the demanded torque quickly. In such

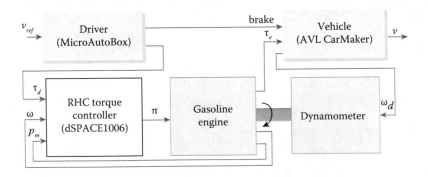

FIGURE 5.10
Block diagram of engine-in-the-loop experiment for torque tracking.

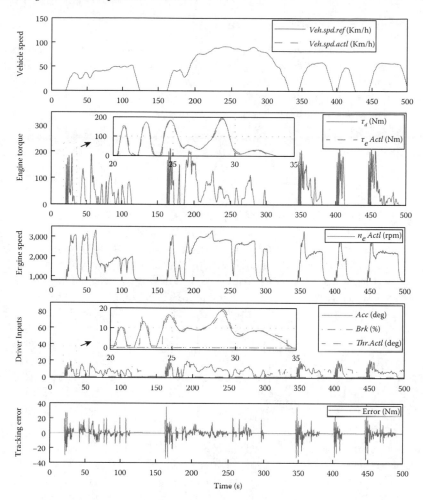

FIGURE 5.11

Engine-in-the-loop experimental results for torque tracking controller.

a closed-loop system, the engine speed $n_e Actl$ varies in real time, which is represented in the third graph. The accelerator position signal Acc, brake position signal Brk, and actual throttle angle $Thr.Actl$ are shown in the fourth graph. The bottom graph illustrates the tracking error between the demanded torque and actual engine torque, which is generally limited in the range of 40 Nm even in the transient process. In general, the control results show the robustness and real-time performance for the proposed RHC torque controller. Because the engine-in-the-loop simulation system provides an approximate engine operating environment, the experimental result indicates the practical engineering significance.

5.7.2   Speed Transient Control

The speed tracking controller is verified in two kinds of different scenarios: idling speed regulation control and wide-range speed tracking control. In order to benchmark the proposed RHC speed tracking controller, the conventional PID control scheme is also implemented in both speed control cases. The PID control calculates the throttle angle according to the speed error between the actual engine speed and the desired command.

Case I: The engine speed is expected to be adjusted around the lower desired speed to reduce the fuel consumption and ensure some accessories, such as air condition and power steering system in normal operations. As for the RHC speed controller, the predictive horizon length is chosen to be 1 s and there are 100 steps over the predictive horizon with the RHC optimization update period of 0.01 s. Besides, the control period is also 0.01 s. The integral gain of the tracking error k_s in the RHC controller is set as -30, and the constraint on the control input is $\vartheta \in [-5,000, 5,000]$. Besides, the weighting coefficients in the performance index are selected as follows: $r_{s1} = r_{s2} = 1$, $r_{s3} = 0.05$. A set of well-tuned design parameters for the PID scheme are chosen as follows: proportional gain $K_p = 0.005$, integral gain $K_i = 0.003$, and differential gain $K_d = 0.001$.

The experimental result for idle speed regulation control is as shown in Figure 5.12. In the experiment, the external load torque follows a step change between 12 and 26 Nm, with the assumption that the air condition is working. Obviously, the RHC speed control is better than the PID control owing to the shorter adjusting time and lower overshoot.

Case II: For hybrid powertrains, the engine speed is usually controlled in a wide operating range to coordinate with the generator torque or gearshift control strategy. In contrast to the former case, the tracking control is more difficult because the obvious system nonlinearity and model error will be revealed in such wide operating conditions. To verify the effectiveness of the proposed RHC speed tracking controller, we conduct the experiments with a sequence of typical test commands, including ramp and step signals in a wide range.

In this case, the fixed integral gain is chosen as -10 and the constraint is $\vartheta \in [-1,000, 1,000]$. Other RHC parameters, such as the predictive horizon length, control period, and weighting coefficients, are the same as in case I. In addition, the PID parameters are chosen as follows: $K_p = 0.01$, $K_i = 0.008$, and $K_d = 0.001$.

Figure 5.13 shows the overall comparative experimental results for speed tracking control with PID and RHC controllers. In the experiments, the dynamometer works under the torque control mode, and thus it can provide the desired load torque. The load torque follows a random step change from 20 Nm to 80 Nm.

From the comparative experiments, we can observe that the actual speed responses by using the PID and RHC control schemes both exhibit good tracking performance even in wide operating ranges. The tracking errors are

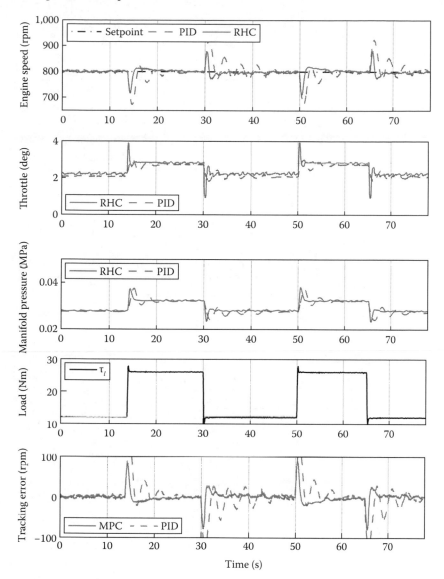

FIGURE 5.12
Experimental results for idle speed regulation control.

constrained under 400 rpm in the fast step process. Meanwhile, it should be noted that the control results with the RHC scheme demonstrate good robustness in the wide operating range. In fact, it benefits from the design of the embedded integrator in the predictive model, which is an effective way to reject the external disturbance and reduce the tracking error.

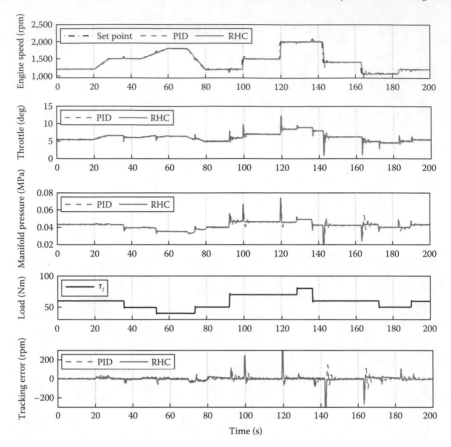

FIGURE 5.13
Experimental results for speed tracking control in wide operating range.

For clearly showing the transient control performance, we give the partial enlarged details during the ramp and step test process, as shown in Figure 5.14. Generally, the performances are similar with respect to the control rapidity between the PID and RHC controller. However, the actual engine speed response with the PID control can easily perform a larger overshoot and a longer settling time than with the RHC control. Essentially, the PID control can be deemed as a one-step-ahead predictive control owing to its differential action, but the RHC can forecast more steps in the future process. And the control action by RHC will be integrated with more future state prediction to guarantee the smoothness of control. Therefore, RHC control performs a better transient tracking performance.

Besides, as is well known, some important RHC parameters, such as predictive horizon length and weighting coefficients, always cause great impact to

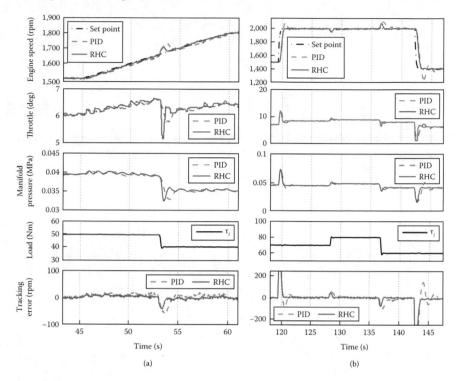

FIGURE 5.14

Comparison of transient control performance between PID and RHC schemes.
(a) Enlarged figure during the ramp command tests. (b) Enlarged figure in
the step command tests.

control performance. To evaluate the control effects with different controller
parameters, some investigations were conducted. The comparative results are
illustrated in Figure 5.15.

From these results, we can see that the parameters should be tuned in
some proper values. For example, in most cases, the predictive horizon should
be close to the time constant of the system that involves the whole transient
process. Too long or too short of a predictive horizon length cannot lead to
good transient control results, as shown in Figure 5.15a. Other important
parameters are the weighting coefficients on the control input. In general, the
smaller weightings on control inputs will lead to a fast control response, but
they will need more control energy, as shown in Figure 5.15b.

It also should be pointed out, in the above control applications, that the
continuation/GMRES method provides approximate optimal solutions with
satisfactory control precision, and real-time performance as well as a low computational burden. Besides, from the experiments, we find that the on-line
optimization algorithm is sensitive to the constraint setup and integral gain.

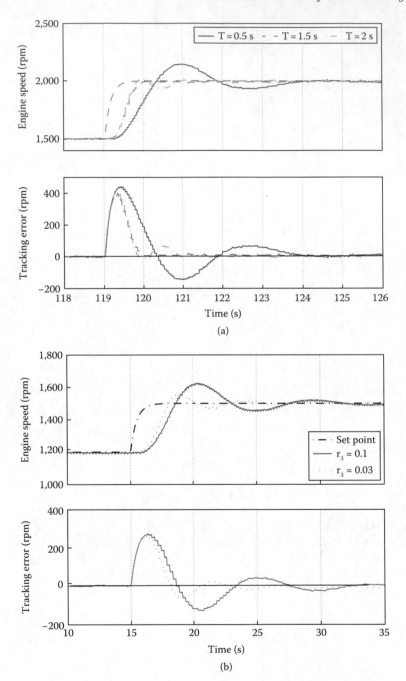

FIGURE 5.15

The impacts of RHC parameters. (a) Impacts of predictive horizon length.
(b) Impacts of weightings on control input.

Therefore, these parameters need to be tuned in practical applications to guarantee the stability of the control system. In fact, the stability issue is still a theoretical challenge for receding horizon control, so we will omit analysis about stability in this study.

5.7.3 Parameter Tuning and Adaptive Compensation

In this section, we verify the parameter tuning approach as discussed in Section 5.5. Recall the aforementioned parameter tuning design for speed tracking control; the weighting coefficients P, Q, and R are the functions with the only adjusting parameter σ. That is, the control performance will be determined by the single parameter.

In order to verify the control performance with the proposed tuning method, we choose three values of the tuning parameter σ, and based on the different values, the RHC weights are shown in Table 5.1.

Then, experimental validations are conducted based on the above weightings choices. Similarly to the previous speed control experiments, the predictive horizon length is set as 1 s and the control period as 0.01 s, and the reference speed command consists of a series of typical components, such as ramp and step signals, to test the transient control performance in a wide operating range. In this case study, the load torque is fixed at 60 Nm. The control performances with different tuning parameter values are compared in Figure 5.16.

It is clearly shown that there are big tracking errors in the whole control process with different tuning parameter values, and the tracking errors vary based on the different operating conditions. Actually, this tracking error is mainly caused by the insufficient model parameters or external disturbance since RIIC is a model-based control method that strongly depends on the model precision.

However, we can observe that the control performances with the different tuning parameter values can be easily compared from these experimental results. The control effects of the step response are enlarged in Figure 5.16b.

TABLE 5.1

RHC weightings with different tuning parameters σ

	$\sigma = 2.6$	$\sigma = 4$	$\sigma = 5$
P	$\begin{bmatrix} 0.8111 & 0.3269 \\ 0.3269 & 0.2600 \end{bmatrix}$	$\begin{bmatrix} 1.0325 & 0.5030 \\ 0.5030 & 0.4000 \end{bmatrix}$	$\begin{bmatrix} 1.1906 & 0.6287 \\ 0.6287 & 0.5000 \end{bmatrix}$
Q	$\begin{bmatrix} 1.0688 & 0.0390 \\ 0.0390 & 0.0221 \end{bmatrix}$	$\begin{bmatrix} 2.5298 & 0.9794 \\ 0.9794 & 0.5941 \end{bmatrix}$	$\begin{bmatrix} 3.9528 & 1.9530 \\ 1.9530 & 1.2426 \end{bmatrix}$
R	0.1	0.1	0.1

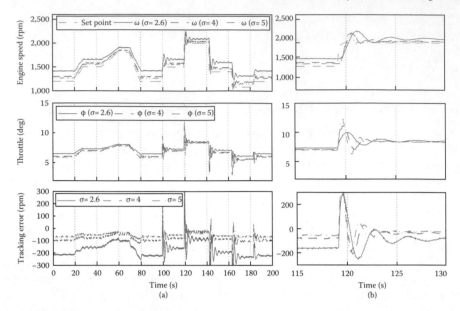

FIGURE 5.16
Experimental results with different tuning parameters σ. (a) Overall experimental results. (b) Transient control process.

The results illustrate that the response speed of the control system is sensitive to the value of the tuning parameter σ. For instance, if the tuning parameter is chosen to be a larger value, the control response will be faster, and accordingly, the magnitude of the control input will be larger. That is, the RHC control performance can be simply adjusted by the proposed tuning parameter. Therefore, the result proves the effectiveness of the proposed tuning approach for the RHC performance.

To compensate for the tracking error, we apply the adaptive controller introduced in Section 5.6 in the above speed tracking experiment. The overall control performances are shown in Figure 5.17. Obviously, with the adaptive controller, the speed tracking errors during the steady conditions are essentially eliminated and the dynamics of the introduced adaptive parameters $\hat{\epsilon}$ are shown in the bottom graph.

Based on this experimental result, the control performances with the different tuning parameter values can be further comparatively investigated (they as are enlarged in Figure 5.18). The comparative results illustrate the control performances of the closed-loop system are distinct from each other and can be adjusted by the different tuning parameters.

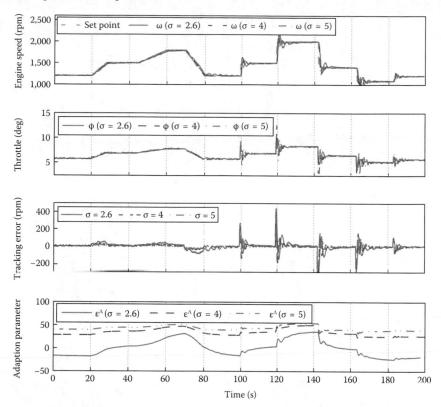

FIGURE 5.17
Overall control performances with adaptive compensation.

5.8 Conclusions

For dynamical systems, receding horizon optimal control is a well-known control strategy to achieve a good transient performance since the control strategy provides on-line optimization subject to the dynamical model and the constraints on the state and the control inputs. However, a bottleneck in the RHC control application is the heavy on-line computation load, especially for the faster systems such as engine control. This chapter challenged the control issues in the engine torque and speed tracking with RHC and provided full-scale experimental validations conducted on a car with a gasoline engine. Also, we investigated the influences of the weightings in the cost function and discussed the parameter tuning method based on the ILQ regulator design.

First, the control-oriented model to match each control issue was exploited with physical observations. For engine torque tracking control, the intake air

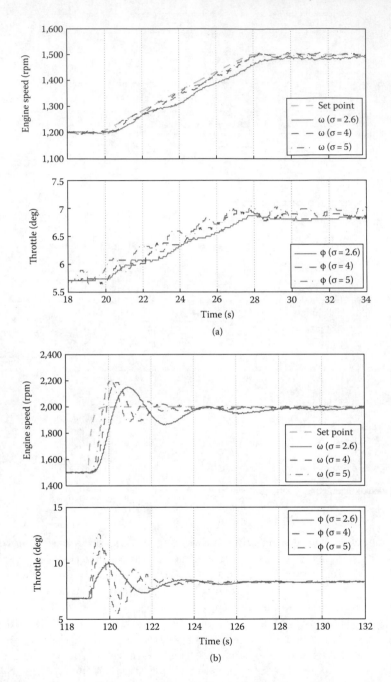

FIGURE 5.18
Comparative results of control performances with different tuning gains.
(a) Ramp responses. (b) Step responses.

charging dynamics was mainly adopted and the rotational dynamics of the crankshaft was ignored due to its very slow response in comparison with torque generation. Then, by means of the air charging dynamical model, the receding horizon controller was designed in the framework of the given continuation/GMRES on-line algorithm. For the engine speed tracking control, the tracking error dynamical system was derived from the engine mean-value model by means of nonlinear per-compensation and coordinate change techniques.

Second, it should be noted that the receding horizon optimization problem was solved based on the model. Therefore, the control performance is usually influenced by accuracy of the model parameters. For combustion engines, the parameters will change the value according to the operation thermal condition and the load variations. Therefore, in practical applications, the integrator of the tracking error is embedded in the predicted model for both speed and torque tracking controllers, and it can effectively improve the control performance.

Third, the on-line optimization of the RHC was achieved by the continuation/GMRES algorithm. In the experimental validations, it has been confirmed that the algorithm has fast optimization capability with lower hardware requirements. Overall, the proposed RHC tracking controllers are able to handle the nonlinear control problem very well with satisfactory real-time performance.

Besides, the proposed parameter tuning approach provides a very convenient tuning parameter for the RHC weightings instead of the traditional cut-and-trial method, so that the control performance of the closed-loop system can be tuned by this single parameter. With the help of this single tuning parameter, we can easily obtain a trade-off between the response speed of the control system and the magnitude of the control input.

6

Balancing Control

6.1 Introduction

In internal combustion engines with multicylinders, each cylinder must generate an equal amount of torque. Furthermore, at a static operating mode, each combustion event in a cylinder must be performed in the same combustion quality. If this is not the case, then an imbalance will occur, and the engine will stumble and lose the smoothness in the power generation. The imbalance is due to the imbalances in the air intake path and the fueling path, the production errors among the cylinders, and the stochastic characteristics of the combustion phenomenon. This chapter deals with the cylinder-to-cylinder balancing problems via feedback control.

6.2 Exhaust Gas Mixing Model

The challenge of the A/F balancing problem is the constraint from the single sensor [87]. Generally, in multicylinder engines, several cylinders share a common exhaust manifold, and the λ-sensor for on-line detection of the A/F is usually equipped at the gas mixing point of the manifold where the exhaust gases from different cylinders come together and flow to the tailpipe.

A feedback control system for a six-cylinder engine is sketched in Figure 6.1, where cylinders 1, 3, and 5 share the exhaust manifold L, and cylinders 2, 4, and 6 share the exhaust manifold R. At the gas mixing point of each exhaust manifold, a universal exhaust gas oxygen (UEGO) sensor is mounted and the fuel mass injection per stroke is decided based on the sensor output. For the six-cylinder engine, since the combustion and the exhaust events of each cylinder occur per 120° along the crank angle, the exhaust valves of the individual cylinders that share the same manifold are opened at intervals of 240°.

In order to represent the exhaust gas mixing phenomenon, the gas mass flow is simply modeled as a response to the first-order linear system under the

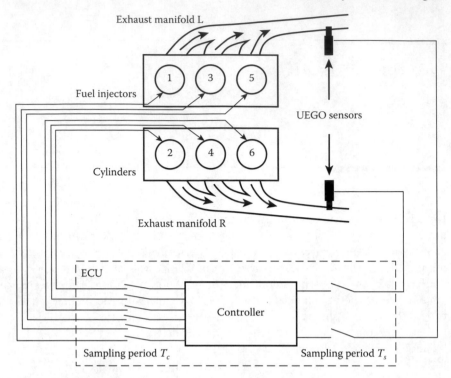

FIGURE 6.1
A/F feedback control system.

pulse excitation, and the information of the A/F is propagated with the gas mass flow. The idea is sketched in Figure 6.2.

Suppose that the components of the exhaust gas are fuel and fresh air. Denote the masses of the fuel and fresh air involved in the exhaust gas by G_f and G_a, respectively. Then, as an equivalent concept of the A/F, the fuel–gas ratio (F/G) is defined by G_f/G_{ex}, where G_{ex} represents the total exhaust gas mass given by

$$G_{ex} = G_a + G_f$$

Then it is easy to show that

$$\frac{G_f}{G_{ex}} = \frac{1}{1+\lambda} \tag{6.1}$$

where $\lambda = G_a/G_f$ denotes the A/F. To normalize the F/G with the theoretical A/F $\lambda_d = 14.7$, we define

$$\eta = (1+\lambda_d)\frac{G_f}{G_{ex}} = \frac{1+\lambda_d}{1+\lambda} \tag{6.2}$$

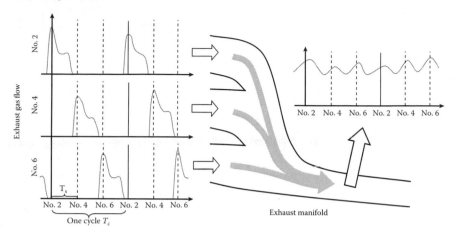

FIGURE 6.2
Exhaust gas flow of each cylinder.

Obviously, $\eta = 1$ when λ takes the value 14.7. Consider cylinders 2, 4, and 6, which share the exhaust manifold R. The following gives a description for the F/G in the gas mixing point.

For the sake of simplicity, we suppose that the exhaust valve of each cylinder will open at the corresponding bottom dead center (BDC). Then as mentioned previously, the mass flow $\dot{m}_{s,i}$ traveling in the exhaust port to the gas mixing point can be represented as

$$\dot{m}_{s,i} = G_{s,i}(s)\dot{m}_{e,i}, \quad G_{s,i}(s) = \frac{1}{\tau_i s + 1} \tag{6.3}$$

where τ_i $(i = 2, 4, 6)$ denotes time constants that are determined by the geometric shape of the exhaust runners.

Furthermore, suppose that the A/F of each cylinder is constant during one cycle. Then at time t, the exhaust gas flow observed at the gas mixing point is $\dot{m}_{s,2}(t) + \dot{m}_{s,4}(t) + \dot{m}_{s,6}(t)$, and the total fuel mass contained in the gas flow is $\eta_2 \dot{m}_{s,2}(t) + \eta_4 \dot{m}_{s,4}(t) + \eta_6 \dot{m}_{s,6}(t)$, where η_i $(i = 2, 4, 6)$ denotes the gas–fuel ratio (G/F) of the ith cylinder. Thus, the G/F observed at the gas mixing point is given by

$$\eta(t) = \frac{\eta_2 \dot{m}_{s,2}(t) + \eta_4 \dot{m}_{s,4}(t) + \eta_6 \dot{m}_{s,6}(t)}{\dot{m}_{s,2}(t) + \dot{m}_{s,4}(t) + \dot{m}_{s,6}(t)} \tag{6.4}$$

From physical consideration, suppose that in the exhaust manifold, the exhaust gas does not remain in the next cycle, and the profile of $\dot{m}_{si}(\tau)$ is repeated cyclically at an ideal static operation mode; that is, if we reset the initial time $\tau = 0$ at each BDC (the starting timing of the exhaust phase), then the gas flow from the ith cylinder passing through the mixing point is represented by

$$\dot{m}_{s,i}(\tau) = \phi_i(\tau) = \begin{cases} \int_0^\tau e^{-\frac{\tau-s}{\tau_i}} \dot{m}_{e,i}(s)ds, & \tau < T_c \\ 0, & \tau \geq T_c \end{cases} \qquad (6.5)$$

where $T_c = 4\pi/\omega$ is the period of time occupied by one engine cycle when the engine speed is ω (rad/s).

Furthermore, note that the exhaust events sequentially appear with time period T_s. Thus, in the time domain t $(0 \leq t < \infty)$, the behavior of the gas flow passing through the mixing point can be represented as a periodic function defined for the ith cylinder by

$$\phi_{e,i}(t) = \phi_i(t - (kT_c + (r-1)T_s)) \qquad (6.6)$$
$$t \in [kT_c + (r-1)T_s, (k+1)T_c + (r-1)T_s]$$
$$k = 0, 1, \cdots, \infty, \quad r = 1, 2, \cdots, N$$

On the other side, according to (6.4), we obtain the following model of the F/G in the gas mixing point:

$$\eta(t) = \frac{\eta_2\phi_{e,2}(t) + \eta_4\phi_{e,4}(t) + \eta_6\phi_{e,6}(t)}{\phi_{e,2}(t) + \phi_{e,4}(t) + \phi_{e,6}(t)} \qquad (6.7)$$

6.3 Individual Cylinder A/F

In this section, the goal is the establishment of an individual cylinder A/F model for the feedback balancing control. As shown in Figure 6.2, the combusted gas of each cylinder is discharged into each exhaust port in the exhaust BDC and flows to the gas mixing point of the exhaust manifold. Now define the F/G in the exhaust port right after the exhaust BDC of the ith cylinder as $\eta_i(k)$ $(i = 2, 4, 6)$ in the kth cycle and let $\eta(k)$ denote the F/G in the gas mixing point of the kth cycle. Since η_i is propagated to the gas mixing point from each exhaust path, if we focus on the individual exhaust gas flow $\dot{m}_{ei}$ and the exhaust flow at the gas mixing point, then the following modeling of the individual cylinder F/G can be done.

6.3.1 Individual Cylinder Fuel–Gas Ratio Modeling

Take the exhaust BDC timing of cylinder 2 as the starting time of T_s. Then three sampling data for $\eta(k)$ will be obtained at $kT_c + rT_s$ $(r = 1, 2, 3)$. Denote the corresponding data $\eta(k)$ at $kT_c + rT_s$ as $\eta^{(r)}(k)$. Since $\eta_i(t)$ is constant during each cycle, the following expression can be obtained from (6.7):

$$\eta^{(r)}(k) = \eta(kT + (r-1)T_s)$$
$$= \gamma_{r,2}(k)\eta_2(k) + \gamma_{r,4}(k)\eta_4(k) + \gamma_{r,6}(k)\eta_6(k) \qquad (6.8)$$

where $\gamma_{r,i}(k)$ $(i = 2, 4, 6)$ is calculated by

$$\gamma_{r,i}(k) = \frac{\phi_{e,i}(kT_c + (r-1)T_s)}{\sum\limits_{j=2,4,6} \phi_{e,j}(kT_c + (r-1)T_s)} \qquad (6.9)$$

We now introduce the following relation with a matrix $\Gamma(k)$,

$$\begin{bmatrix} \eta^{(1)}(k) \\ \eta^{(2)}(k) \\ \eta^{(3)}(k) \end{bmatrix} = \Gamma(k) \begin{bmatrix} \eta_1(k) \\ \eta_2(k) \\ \eta_3(k) \end{bmatrix} \qquad (6.10)$$

and furthermore, if matrix $\Gamma(k)$ is invertible, denote $[\Gamma(k)]^{-1}$ as the following matrix,

$$[\Gamma(k)]^{-1} = \begin{bmatrix} \sigma_{21}(k) & \sigma_{22}(k) & \sigma_{23}(k) \\ \sigma_{41}(k) & \sigma_{42}(k) & \sigma_{43}(k) \\ \sigma_{61}(k) & \sigma_{62}(k) & \sigma_{63}(k) \end{bmatrix} \qquad (6.11)$$

by which the following equation can be obtained:

$$\eta_i(k) = \sigma_{i1}(k)\eta^{(1)}(k) + \sigma_{i2}(k)\eta^{(2)}(k) + \sigma_{i3}(k)\eta^{(3)}(k) \qquad (6.12)$$

Furthermore, to deal with the modeling based on the measurements from the sensor, it is essential to integrate the sensor characteristics into the modeling. Considering the delay caused by the sensor, define a "filtered" F/G of the individual $\eta_{s,i}$ as

$$\eta_{s,i} = G_{\text{sen}}(s)\eta_i \qquad (6.13)$$

with the following transfer function:

$$G_{\text{sen}}(s) = \frac{e^{-\tau_d s}}{\tau_s s + 1} \qquad (6.14)$$

Taking the z-transformation of (6.14), we obtain

$$G_{\text{sen}}(z) = \frac{g}{z^m(z - c)} \qquad (6.15)$$

where $g = 1 - e^{-(T_c/\tau_s)}$ and $c = e^{-(T_c/\tau_s)}$. For the sake of simplicity, we assume that $\tau_d = mT_c$ but m is unknown. With (6.15), the following equations can be shown from (6.13):

$$z^{-(m+1)}g\eta_i(k) = \eta_{s,i}(k) - c\eta_{s,i}(k-1) \tag{6.16}$$

$$z^{-(m+1)}g\eta(k) = \eta_s(k) - c\eta_s(k-1) \tag{6.17}$$

Considering (6.12) with (6.16) and (6.17) gives the following individual η_{si} representation that involves the sensor characteristic:

$$\eta_{s,i}(k) = c\eta_{s,i}(k-1) + \sum_{r=1}^{3}\sigma_{ir}\eta_s^{(r)}(k) - c\sum_{r=1}^{3}\sigma_{ir}\eta_s^{(r)}(k-1) \tag{6.18}$$

Taking into account the error $w_i(k)$ in the modeling, such as modeling uncertainties and measurement noise, the model (6.18) can be rewritten as

$$\eta_{s,i}(k) = \varphi_i^T(k)\theta_i + w_i(k) \tag{6.19}$$

with the parameter vector θ_i and the function $\varphi_i(k)$ defined as

$$\theta_i = \begin{bmatrix} c \\ \sigma_{i1} \\ \sigma_{i2} \\ \sigma_{i3} \\ -c\sigma_{i1} \\ -c\sigma_{i2} \\ -c\sigma_{i3} \end{bmatrix}, \quad \varphi_i(k) = \begin{bmatrix} \eta_{si}(k-1) \\ \eta_s^{(1)}(k) \\ \eta_s^{(2)}(k) \\ \eta_s^{(3)}(k) \\ \eta_s^{(1)}(k-1) \\ \eta_s^{(2)}(k-1) \\ \eta_s^{(3)}(k-1) \end{bmatrix}$$

Finally, it can be noted that a linear time-invariant model is derived for the estimation of the individual cylinder F/G.

6.3.2 Model Validation

Validation of the model (6.19) is performed with the engine test bench shown in Chapter 1. Note that under the same operating conditions, the parameters θ_i of the model (6.19) can be considered constant since ϕ_i is a period function. However, for an engine system, it is impractical to get the parameters θ_i of the model (6.19) by theoretical analysis. Experimental data obtained from the engine test bench are used to identify the model parameters. As shown in Figure 6.3a, from the installed UEGO sensors, the sampled data $\eta_{s,i}(k)$ ($i = 2, 4, 6$) and $\eta_s^r(k)$ ($r = 1, 2, 3$) can be obtained as shown in Figure 6.3b. Identification experiments are conducted after warming up the engine and under static operation modes with respect to the engine speed and the external load. Delivering the individual fuel injection commands with 2% steps, the

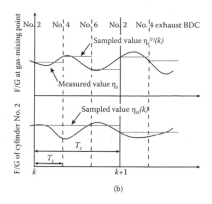

(a)

(b)

FIGURE 6.3

(a) Engine test bench with individual UEGO sensors. (b) Image of the F/G sampling period.

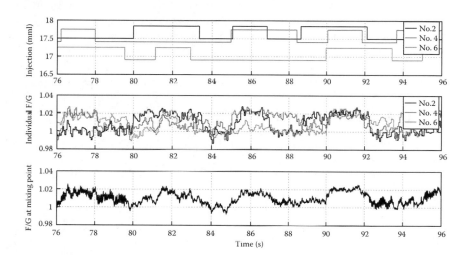

FIGURE 6.4

Experimental data for parameter identification: engine speed = 2,000 rpm, load = −400 mmHg.

measured data used for identification are as shown in Figure 6.4. Then the RLS algorithm is applied to get the estimation of the parameters $\hat{\theta}_i$. The identification results are shown in Table 6.1.

A validation result obtained under the same operating mode as in model (6.19), with the identified parameters against the experimental data, is shown in Figure 6.5. This result shows that the model can give an accurate estimation of the individual F/G.

TABLE 6.1

Identified parameters: engine speed $= 2{,}000$ rpm,
load $= -400$ mmHg

	$i = 2$	$i = 4$	$i = 6$
$\hat{\theta}_{i1}$	0.784	0.745	0.726
$\hat{\theta}_{i2}$	-2.203	1.084	0.920
$\hat{\theta}_{i3}$	1.233	-2.545	-0.786
$\hat{\theta}_{i4}$	1.243	2.611	1.315
$\hat{\theta}_{i5}$	0.509	-0.376	-0.514
$\hat{\theta}_{i6}$	-0.211	1.194	1.483
$\hat{\theta}_{i7}$	-0.353	-1.712	-2.144

Furthermore, the model (6.19) is developed under the assumption that the engine is at static operating mode. To show the parameter variation according to the operating modes, identifications are performed for nine operating modes. Table 6.2 lists the variances of the errors between $\eta_{s,i}(k)$ and $\hat{\eta}_{s,i}(k)$, which are obtained with the corresponding identified parameters. It can be observed that the variances are tolerant levels.

6.4 A/F Balancing Control

Based on the above estimation method, the individual A/F balancing control problem will be investigated in this section. The goal of the balancing control is to regulate the individual fuel injection $u_{f,i}$ such that the filtered individual F/G $\eta_{s,i}$ is the desired value $\eta_d = 1$.

Taking into account the fuel path dynamics and the transfer characteristics of the filter (the virtual sensor), the dynamics from the fuel injection command of the ith cylinder $u_{f,i}$ to the output of the fuel path $\eta_{s,i}$ can be represented by the following model:

$$A_i(z^{-1})\eta_{s,i}(k) = z^{-d}B_i(z^{-1})u_{f,i}(k) + C_i(z^{-1})w_{n,i}(k) \qquad (6.20)$$

$$A_i(z^{-1}) = 1 + a_{1i}z^{-1} + a_{2i}z^{-2}$$

$$B_i(z^{-1}) = b_{0i}$$

$$C_i(z^{-1}) = 1$$

$$d = 1$$

where w_{ni} denotes the white noise, including the error in the model. Denote the model parameter vector as $\theta_{ci} = [a_{1i}\ a_{2i}\ b_{0i}]^T$.

A self-tuning regulator and an adaptive generalized predictive control (GPC) will be shown below. Taking the V-type six-cylinder SI engine as

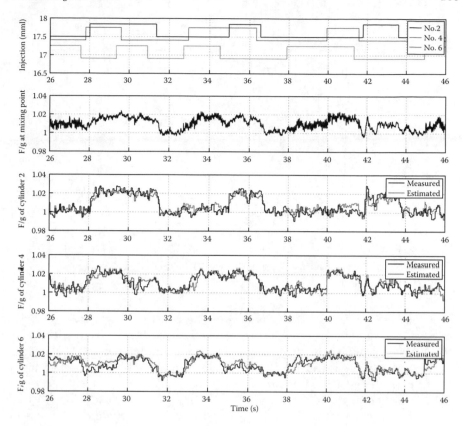

FIGURE 6.5
Validation of individual F/G estimation model: engine speed = 2,000 rpm, load = −400 mmHg.

TABLE 6.2
Variances of $\hat{\eta}_{si}$ estimation error $(\times 10^{-5})$ under different operating modes

Speed (rpm)	Load (mmHg)	$i = 2$	$i = 4$	$i = 6$
1,600	−400	1.331	0.825	0.735
	−300	2.757	2.505	1.971
	−200	2.714	2.694	1.792
2,000	−400	2.190	1.594	1.069
	−300	3.140	2.371	2.062
	−200	3.442	3.154	2.535
2,400	−400	3.516	2.742	1.339
	−300	5.209	3.415	1.922
	−200	7.990	6.633	3.269

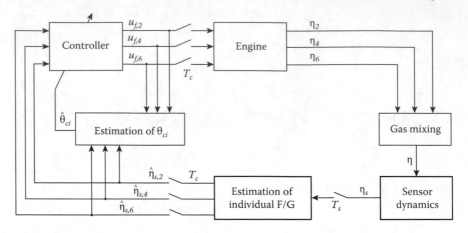

FIGURE 6.6
Outline of the individual F/G control system.

an example $(i = 2, 4, 6)$, the structure of the control system is as shown in Figure 6.6, where it can be seen that for the control design, the model parameter θ_{ci} is estimated online with the estimated individual F/G. Then, the self-tuning regulator and GPC algorithms are derived based on the model (6.20) with the estimated parameter $\hat{\theta}_{ci}$.

As shown in Figure 6.6, the control law will be solved at each cycle with the estimated parameter $\hat{\theta}_{ci}$, which is obtained from the RLS algorithm by using the estimated $\hat{\eta}_{s,i}(k)$. A remark on the control scheme is that it was developed for proper static operating modes, since the model for providing $\hat{\eta}_i(k)$ is effective at the corresponding operating modes. By resetting the parameters of the model (6.19), the control scheme might adapt the operating modes.

6.4.1 Self-Tuning Regulator

As mentioned before, since the parameters of the model (6.20) are identified first, an indirect self-tuning regulator is employed. The self-tuning regulator design problem is solved by the minimum-variance control. Due to the time delay, control input $u_{fi}(k)$ can only contribute to the output at $k + d$. Hence, the index function is chosen as follows:

$$J = E[(\eta_{s,i}(k+d) - \eta_d(k+d))^2] \qquad (6.21)$$

where $E[\cdot]$ denotes the expectation operator and $\eta_d(= 1)$ is the set-point desired output. Finding the control $u_{fi}(k)$ that minimizes J (i.e., $u_{f,i}(k) = \arg\min J(u_{f,i})$) needs to predict the output $\eta_{s,i}(k+d)$ at k. The following shows the details of the predication and the self-tuning regulator design.

From the model (6.20), it can be obtained that

$$\eta_{s,i}(k+d) = \frac{B_i(z^{-1})}{A_i(z^{-1})}u_{f,i}(k) + \frac{C_i(z^{-1})}{A(z^{-1})}w_{n,i}(k+d) \tag{6.22}$$

Applying the solutions $R_i(z^{-1}) = 1$ and $H_i(z^{-1}) = -a_{1i} - a_{2i}z^{-1}$ of the equation $C_i(z^{-1}) = A_i(z^{-1})R_i(z^{-1}) + z^{-d}H_i(z^{-1})$ to the above expression gives

$$\eta_{s,i}(k+d) = \frac{B_i(z^{-1})}{A(z^{-1})}u_{f,i}(k) + \frac{H_i(z^{-1})}{A_i(z^{-1})}w_{n,i}(k) + R_i(z^{-1})w_{n,i}(k+d) \tag{6.23}$$

Furthermore, by using the relationship of Equation 6.20, the expression (6.23) becomes

$$\eta_{s,i}(k+d) = \frac{B_i(z^{-1})R_i(z^{-1})}{C_i(z^{-1})}u_{f,i}(k) + \frac{H_i(z^{-1})}{C_i(z^{-1})}\eta_{s,i}(k) + R_i(z^{-1})w_{n,i}(k+d) \tag{6.24}$$

Denote $\bar{\eta}_{s,i}(k+d|k)$ as the predicted value of $\eta_{s,i}(k+d)$ at k. Then, this value is achieved in the sense of minimum-variance prediction. The objective is to design a predictor to generate the $\bar{\eta}_{s,i}(k+d|k)$ that minimizes the following expectation:

$$E\left[(\bar{\eta}_{s,i}(k+d|k) - \eta_{s,i}(k+d))^2\right] \tag{6.25}$$

Substituting (6.24) into (6.25) and considering the relation that $E[(R_i(z^{-1})w_{n,i}(k+d))^2]$ is independent of the choice of $\bar{\eta}_{s,i}(k+d|k)$, we obtain the prediction that minimizes (6.25):

$$\bar{\eta}_{s,i}(k+d|k) = \frac{B_i(z^{-1})R_i(z^{-1})}{C_i(z^{-1})}u_{f,i}(k) + \frac{H_i(z^{-1})}{C_i(z^{-1})}\eta_{s,i}(k) \tag{6.26}$$

With the above output prediction, the index function (6.21) can be written as

$$J = E\left[(\hat{\eta}_{s,i}(k+d|k) + \tilde{\eta}_{s,i}(k+d|k) - \eta_d(k+d))^2\right]$$

Note that the predicting error $\tilde{\eta}_{s,i}(k+d|k)$ is a linear combination of $w_{n,i}(k+d)$; hence, the necessary and sufficient condition for minimizing the index function is

$$\bar{\eta}_{s,i}(k+d|k) - \eta_d(k+d) = 0 \tag{6.27}$$

From the prediction (6.26), the above condition gives the control law as

$$u_{f,i}(k) = \frac{1}{B_i(z^{-1})R_i(z^{-1})}(H_i(z^{-1})\eta_{s,i}(k) + C_i(z^{-1})\eta_d(k+d)) \tag{6.28}$$

By inserting the coefficients of the model (6.28), the control law can be arranged as

$$u_{f,i}(k) = \frac{1}{b_{0i}}[\eta_d(k+1) + a_{1i}\eta_{s,i}(k) + a_{2i}\eta_{s,i}(k-1)] \tag{6.29}$$

To implement the control law (6.29), $\eta_{s,i}(k)$ and $\eta_{s,i}(k-1)$, which cannot be measurable for feedback control, are replaced by the estimations $\bar{\eta}_{s,i}(k)$ and $\bar{\eta}_{s,i}(k-1)$ obtained from the proposed individual F/G estimation algorithm. A validation result under the operating mode with an engine speed and load of 1,600 rpm and -400 mmHg, respectively, is as shown in Figure 6.7. In the validating experiment, the fuel injections to each cylinder are provided by a PI controller using the F/G measured at the gas mixing point initially, and then, at $t = 46$ s, the individual fuel injection is activated by using the self-tuning regulator with the estimation. From this point, the result shows that

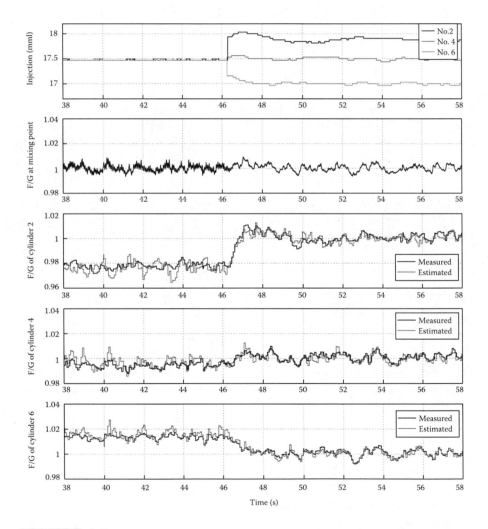

FIGURE 6.7
Individual F/G control result using self-tuning regulator with estimation: engine speed $= 1,600$ rpm, load $= -400$ mmHg.

the fuel injection for each cylinder is individually delivered, and both the estimated and measured individual F/G converge to 1. Moreover, the result can demonstrate that the model presented above provides the individual F/G with sufficient precision.

6.4.2 Generalized Predictive Control

Considering that the disturbance term in the system includes a nonstationary element, the model (6.20) is rewritten as the following controlled autoregressive and integrated moving-average (CARIMA) model:

$$A_i(z^{-1})\eta_{s,i}(k) = z^{-1}B_i(z^{-1})u_{f,i}(k) + C_i(z^{-1})\frac{e_i(k)}{\Delta} \tag{6.30}$$

where $\Delta = 1 - z^{-1}$ and e_i denotes a zero-mean white noise. Based on the above model (6.30), the GPC algorithm is derived in the following according to the technique presented in [84, 85].

The objective of the predictive control is to compute the future control sequence $u_{f,i}(k), u_{f,i}(k+1), \cdots$, in a way that the j-step future plant output $\eta_{s,i}(k+j)$ is close to a reference trajectory. This is accomplished by minimizing the cost function given by

$$J_i(N_1, N_2, N_u) = \sum_{j=N_1}^{N_2} \delta(j)[\bar{\eta}_{s,i}(k+j \mid k) - w_{r,i}(k+j)]^2$$

$$+ \sum_{j=1}^{N_u} \lambda(j)[\Delta u_{f,i}(k+j-1)]^2 \tag{6.31}$$

where N_1 and N_2 denote the minimum and maximum prediction horizons, respectively, N_u denotes the control horizon, $\delta(j)$ and $\lambda(j)$ denote the weighting sequences, and $w(k+j)$ denotes the reference trajectory sequence that is calculated by the following first-order lag model:

$$w_{r,i}(k+n) = \alpha w_{r,i}(k+n-1) + (1-\alpha)\eta_d(k+n)$$

$$\text{with } w_{r,i}(k) = \eta_{s,i}(k), \quad \alpha \in [0,1], \quad \eta_d = 1 \text{ and } n = 1 \cdots N_2 \tag{6.32}$$

First, the j-step future output $\eta_{si}(k+j)$ can be derived as

$$\eta_{s,i}(k+j) = F_{j,i}\eta_{s,i}(k) + E_{j,i}(z^{-1})B_i(z^{-1})\Delta u_{f,i}(k+j-1)$$

$$+ E_{j,i}(z^{-1})e_i(k+j) \tag{6.33}$$

where $E_{j,i}$ and $F_{j,i}$ are polynomials satisfying the following Diophantine equation:

$$1 = E_{j,i}(z^{-1})\Delta A_i(z^{-1}) + z^{-j}F_{j,i}(z^{-1}) \tag{6.34}$$

and they are uniquely defined with degree $j-1$ and degree $A_i(z^{-1})$, respectively. In the sense of minimum-variance prediction with respect to the expectation $E[(\bar{\eta}_{s,i}(k+j|k) - \eta_{s,i}(k+j))^2]$, the j-step predictor of the output can be obtained as

$$\bar{\eta}_{s,i}(k+j|k) = G_{j,i}(z^{-1})\Delta u_{fi}(k+j-1) + F_{j,i}(z^{-1})\eta_{s,i}(k) \qquad (6.35)$$

where $G_{j,i}(z^{-1}) = E_{j,i}(z^{-1})B_i(z^{-1})$ and $G_{j,i}$ is represented by the following polynomial:

$$G_{j,i}(z^{-1}) = g_{0,i} + g_{1,i}z^{-1} + \cdots + g_{j,i}z^{-j}$$

For the considered F/G control problem, choose the following values for the horizons and the weightings:

$$N_1 = 1, \quad N_2 = 5, \quad N_u = 5, \quad \delta(j) = 1, \quad \lambda(j) = 0.1$$

Then from the model (6.35), the following sequence of output predications can be obtained:

$$\begin{aligned}
\bar{\eta}_{s,i}(k+1|k) &= G_{1,i}\Delta u_{f,i}(k) + F_{1,i}\eta_{s,i}(k) \\
\bar{\eta}_{s,i}(k+2|k) &= G_{2,i}\Delta u_{f,i}(k+1) + F_{2,i}\eta_{s,i}(k)
\end{aligned}$$

$$\vdots$$

$$\bar{\eta}_{s,i}(k+5\mid k) = G_{5,i}\Delta u_{f,i}(k+4) + F_{5,i}\eta_{s,i}(k)$$

which can be written as

$$\mathbf{y}_i = \mathbf{G}_i\mathbf{u}_i + \mathbf{F}_i(z^{-1})\eta_{s,i}(k) + \mathbf{G}'_i(z^{-1})\Delta u_{f,i}(k-1) \qquad (6.36)$$

where

$$\mathbf{y}_i = \begin{bmatrix} \bar{\eta}_{s,i}(k+1\mid k) \\ \bar{\eta}_{s,i}(k+2\mid k) \\ \vdots \\ \bar{\eta}_{s,i}(k+5\mid k) \end{bmatrix}, \quad \mathbf{u}_i = \begin{bmatrix} \Delta u_{f,i}(k) \\ \Delta u_{f,i}(k+1) \\ \vdots \\ \Delta u_{f,i}(k+4) \end{bmatrix}, \quad \mathbf{G}_i = \begin{bmatrix} g_{0,i} & 0 & \cdots & 0 \\ g_{1,i} & g_{0,i} & \cdots & 0 \\ \vdots & \vdots & \vdots & \vdots \\ g_{4,i} & g_{3,i} & \cdots & g_{0,i} \end{bmatrix}$$

$$\mathbf{F}_i(z^{-1}) = \begin{bmatrix} F_{1,i}(z^{-1}) \\ F_{2,i}(z^{-1}) \\ \vdots \\ F_{5,i}(z^{-1}) \end{bmatrix}, \quad \mathbf{G}'_i(z^{-1}) = \begin{bmatrix} (G_{1,i}(z^{-1}) - g_{0,i})z \\ (G_{2,i}(z^{-1}) - g_{0,i} - g_{1,i}z^{-1})z^2 \\ \vdots \\ (G_{5,i}(z^{-1}) - g_{0,i} - \cdots - g_{4,i}z^4)z^5 \end{bmatrix}$$

Let $\mathbf{f}_i = \mathbf{F}_i(z^{-1})\eta_{si}(k) + \mathbf{G}'_i(z^{-1})\Delta u_{fi}(k-1)$, which are the components dependent on the past in (6.36). The expression for the output predictions can now be given by

$$\mathbf{y}_i = \mathbf{G}_i\mathbf{u}_i + \mathbf{f}_i \qquad (6.37)$$

With this expression, the cost function (6.31) can be written as

$$J_i = (\mathbf{G}_i \mathbf{u}_i + \mathbf{f}_i - \mathbf{w}_{ri})^T (\mathbf{G}_i \mathbf{u}_i + \mathbf{f}_i - \mathbf{w}_{ri}) + \lambda \mathbf{u}_i^T \mathbf{u}_i$$
$$= \frac{1}{2} \mathbf{u}_i^T \mathbf{H}_i \mathbf{u}_i + \mathbf{b}_i^T \mathbf{u}_i + \mathbf{f}_{0i} \tag{6.38}$$

where $\mathbf{w}_{ri} = [w_{ri}(k+1) \ w_{ri}(k+2) \ \cdots \ w_{ri}(k+5)]^T$, $\mathbf{H_i} = 2(\mathbf{G}_i^T \mathbf{G}_i + \lambda \mathbf{I})$, $\mathbf{b}_i^T = 2(\mathbf{f}_i - \mathbf{w}_i)^T \mathbf{G}_i$, and $\mathbf{f}_{0i} = (\mathbf{f}_i - \mathbf{w}_{ri})^T (\mathbf{f}_i - \mathbf{w}_{ri})$.

Assume that there are no constraints on the control signals. The minimization of J_i can be achieved by making $\partial \mathbf{H}_i / \partial \mathbf{u}_i = 0$, and this gives the control law as

$$\mathbf{u}_i = (\mathbf{G}_i^T \mathbf{G}_i + \lambda \mathbf{I})^{-1} \mathbf{G}_i^T (\mathbf{w}_{ri} - \mathbf{f}_i) \tag{6.39}$$

where $\mathbf{u}_i = [\Delta u_{fi}(k) \ \Delta u_{fi}(k+1) \ \cdots \ \Delta u_{fi}(k+N_2-1)]^T$ and $\mathbf{w}_i = [w_i(k+1) \ w_i(k+2) \ \cdots \ w_i(k+N_2)]^T$. At the kth cycle, the first element of the vector $\mathbf{u}_i$ is delivered to the fuel injection actuator of cylinder i. Hence, the control law is given by

$$u_{fi}(k) = u_{fi}(k-1) + \mathbf{K}_i(\mathbf{w}_{ri} - \mathbf{f}_i) \tag{6.40}$$

with gain $\mathbf{K}_i$ determined by the first row of the matrix $(\mathbf{G}_i^T \mathbf{G}_i + \lambda \mathbf{I})^{-1} \mathbf{G}_i^T$.

The scheme with GPC (6.40) is validated on the engine test bench. In real-time control, the values of $\eta_{si}(k)$ in the control law (6.40) are replaced by the estimated $\hat{\eta}_{si}(k)$. The coefficient α in (6.32) is chosen as 0.5. The operating mode with an engine speed and load of 1,600 rpm and -400 mmHg, respectively, is tested and the result shown in Figure 6.8. In the validating experiment, the F/G controller is activated at $t = 256$ s. It can be observed that under the control, the individual F/G can be regulated to the desired value $\eta_d = 1$ with good performance.

Finally, there is a need to remark that the control parameters, including the prediction horizons N_1, N_2, and N_u, the weighting functions $\delta(j)$ and $\lambda(j)$, and α in Equation 6.32, will affect the control performance. In practical engineering, proper numerical optimization methods are recommended for searching better design parameters.

6.5 Linear Time-Varying Model and Offset Learning

For the engines with multicylinders, an estimation approach of the individual F/G is presented by using a linear time-invariant model (6.19) in the previous section, and based on this model, the equation (6.20) represents a multi-input multi-output (MIMO) control system as shown in Figure 6.6. In this section, a periodic time-varying modeling approach proposed in [88] will be introduced. This modeling approach generates a single-input single-output (SISO) model for the fuel path. This modeling technique is then applied to

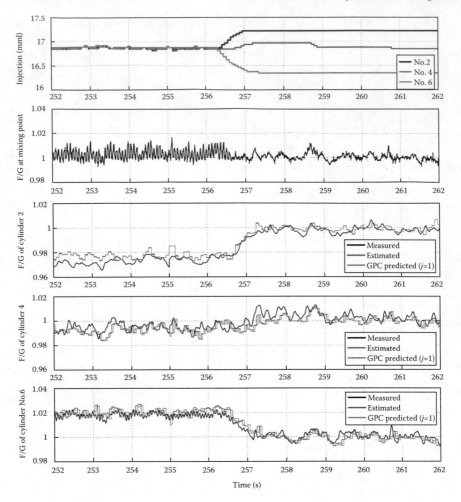

FIGURE 6.8
Individual F/G control results using GPC with estimation: engine speed $= 1{,}600$ rpm, load $= -400$ mmHg.

deal with the issues of state observer design and the learning control of the cylinder-to-cylinder offset.

6.5.1 Periodic Time-Varying Fuel Path Model

As shown in Figure 6.9, the fuel path of the engine with multiple cylinders is actually a multi-input single-output (MISO) system with the individual fuel injection input u_{fi} $(i = 1, 2, \cdots, N)$ and the output F/G η at the gas mixing point of the exhaust system. In the modeling approach presented below,

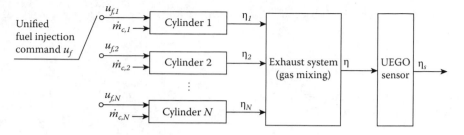

FIGURE 6.9
Schematic of the process from fuel injection to A/F.

a unified fuel injection command u_f that is a periodic time-varying signal is introduced. In this case, the MISO physical system is transferred into a SISO system.

First, consider the continuous time domain. Reviewing Section 6.3.1, the model (6.7) can be rewritten as

$$\eta(t) = \sum_{i=1}^{N} \gamma_i(t)\eta_i(t) \quad \text{with} \quad \gamma_i(t) = \frac{\phi_{e,i}(t)}{\sum_{i=1}^{N} \phi_{e,i}(t)} \tag{6.41}$$

Then, involving the individual $\eta_i(t) = \dot{m}_{f,i}(t)/\dot{m}_{e,i}(t)$ gives

$$\eta(t) = \sum_{i=1}^{N} r_i(t)\dot{m}_{f,i}(t) \quad \text{with} \quad r_i(t) = \frac{\gamma_i(t)}{\dot{m}_{e,i}(t)} \tag{6.42}$$

Furthermore, for the engines using a direct injection system, the injected fuel mass flow to cylinder i can be approximated by the following equation:

$$\dot{m}_{f,i}(t) = u_{f,i}(t) + d_i \tag{6.43}$$

In Equation 6.43, d_i denotes the unknown offset in the fuel path. Then, the output F/G can be described by the following equation:

$$\eta(t) = \sum_{i=1}^{N} r_i(t)(u_{fi}(t) + d_i) \tag{6.44}$$

Finally, taking into account the sensor delay, which is usually described by a linear first-order system with a time constant τ_s, the measured F/G η_s can be obtained by

$$\dot{\eta}_s(t) = \frac{1}{\tau_s}(-\eta_s(t) + \eta(t)) \tag{6.45}$$

Let j denote the index of the exhaust BDC timing of each cylinder. Considering the periodic characteristic of Equation 6.6, at the BDC timing of cylinder i during the kth cycle, the sampling value of a continuous-time signal $y(t)$ can be given by

$$y(j) = y(jT_s) = y(kT + (i-1)T_s) \tag{6.46}$$

Hence, regarding the fuel injection command, the sampling values have the following relationship:

$$u_{f,i}(kN+i-1) = u_{f,i}(kN+i) = \cdots = u_{f,i}((k+1)N+i-2) \tag{6.47}$$

With the above fact in mind, we get the following discrete-time models:

- When $\mathbf{mod}(j,N) = 0$ (i.e., at the BDC of cylinder 1),

$$\eta(j) = r_1(j)(u_{f,1}(j)+d_1) + \sum_{i=2}^{N} r_i(j)(u_{f,i}(j-N+i-1)+d_i) \tag{6.48}$$

- When $\mathbf{mod}(j,N) = p-1$ (i.e., at the BDC of cylinder p $[p < N]$),

$$\eta(j) = \sum_{i=1}^{p} r_i(j)(u_{f,i}(j+i-p)+d_i)$$
$$+ \sum_{i=p+1}^{N} r_i(j)(u_{f,i}(j-N+i-p)+d_i) \tag{6.49}$$

- When $\mathbf{mod}(j,N) = N-1$ (i.e., at the BDC of cylinder N),

$$\eta(j) = \sum_{i=1}^{N} r_i(j)(u_{f,i}(j-N+i)+d_i) \tag{6.50}$$

where the sampled value of $u_{fi}(j)$ is given according to the following expressions:

$$u_{f,i}(j) = \begin{cases} u_{f,i}((k-1)N+i-1), & j < kN+i-1 \\ u_{f,i}(kN+i-1), & j \geq kN+i-1 \end{cases} \tag{6.51}$$

For the sensor dynamics, the discrete-time representation of the differential equation (6.45) is given by

$$\eta_s(j+1) = \alpha\eta_s(j) + (1-\alpha)\eta(j) \quad \text{with} \quad \alpha = 1 - \frac{T_s}{\tau_s} \tag{6.52}$$

Furthermore, we introduce a unified injection command $u_f(j)$. Then, it can be known that $u_f(j) = u_{fi}(j)$ when $\mathbf{mod}(j, N) = i - 1$. To formulate a state-space model, define the state variables as

$$x(j) = \begin{bmatrix} x_1(j) \\ x_2(j) \\ \vdots \\ x_{N-1}(j) \\ x_N(j) \end{bmatrix} = \begin{bmatrix} u_f(j-1) + m(j-1)d \\ u_f(j-2) + m(j-2)d \\ \vdots \\ u_f(j-N+1) + m(j-N+1)d \\ \eta_s(j) \end{bmatrix} \qquad (6.53)$$

where $d = [d_1 \ d_2 \ \cdots \ d_N]^T$ and $m(j) = \beta \Gamma^k$ with

$$\beta = \begin{bmatrix} 1 \\ 0 \\ \vdots \\ 0 \end{bmatrix}_{1 \times N} \qquad \Gamma = \begin{bmatrix} 0 & 1 & 0 & \cdots & & \cdots & 0 \\ 0 & 0 & 1 & 0 & & \cdots & 0 \\ \vdots & & & \ddots & & & \vdots \\ \vdots & & & & \ddots & & 0 \\ 0 & & & & & \ddots & 1 \\ 1 & 0 & \cdots & & \cdots & & 0 \end{bmatrix}_{N \times N}$$

With the above matrices, the models (6.48) through (6.50) can be represented by a unified expression,

$$\eta(j) = \sum_{i=1}^{N} (\beta \Gamma^j R_i(j)) \beta \Gamma^{j-i+1} M(j) \qquad (6.54)$$

where

$$M(j) = \begin{bmatrix} u_f(j) + m(j)d \\ u_f(j-1) + m(j-1)d \\ \vdots \\ u_f(j-N+1) + m(j-N+1)d \end{bmatrix} \qquad R_i(j) = \begin{bmatrix} r_i(j) \\ r_i(j) \\ \vdots \\ r_i(j) \end{bmatrix}$$

With the state definition (6.53), the model (6.54) can be rewritten as

$$\eta(j) = \sum_{i=1}^{N} l_i(j) x_i(j) + l_N(u(j) + m(j)d) \qquad (6.55)$$

where $l_i(j)$ $(i = 1, 2, \cdots, N)$ represents the periodic transfer factor from the fuel mass to the F/G.

By the above description, the following linear periodic time-varying model can be obtained for the fuel path system shown in Figure 6.9:

$$\begin{cases} x(j+1) = A(j)x(j) + B(j)(u(j) + m(j)d) \\ y(j) = Cx(j) \end{cases} \tag{6.56}$$

where

$$A(j) = \begin{bmatrix} 0 & 0 & \cdots & & \cdots & 0 \\ 1 & 0 & \cdots & & \cdots & 0 \\ \vdots & & \ddots & & & \vdots \\ 0 & & \cdots & 1 & 0 & 0 \\ (1-\alpha)l_1(j) & \cdots & & \cdots & (1-\alpha)l_{N-1}(j) & \alpha \end{bmatrix},$$

$$B(j) = \begin{bmatrix} 1 \\ 0 \\ \vdots \\ 0 \\ (1-\alpha)l_N(j) \end{bmatrix}, \quad C = \begin{bmatrix} 0 \\ 0 \\ \vdots \\ 0 \\ 1 \end{bmatrix}^T$$

6.5.2   State Observer

A state observer design method with respect to a linear time-varying system (6.56) in the situation when $d = 0$ is presented in [89]. The following proposes an application to the balancing control of SI engines.

Consider the following SISO linear time-varying system:

$$\begin{cases} x(k+1) = A(k)x(k) + B(k)u(k) \\ y(k) = C(k)x(k) \end{cases} \tag{6.57}$$

where $x \in R^n$, $u \in R^1$, and $y \in R^1$. Introduce the following state transition matrix:

$$\Phi(j_1, j_2) = A(j_1 - 1)A(j_1 - 2) \cdots A(j_2), \quad j_1 > j_2 \tag{6.58}$$

For designing the state observer for the above system, fundamental conceptions with respect to the time-varying system are presented in the following [86].

Definition 6.1   System (6.57) is said to be completely reachable from the origin within n steps if and only if for any $x_1 \in R^n$ there exists a bounded input $u(l)$ $(l = j, \cdots, j+n-1)$ such that $x(0) = 0$ and $x(j+n) = x_1$ for all j.

Definition 6.2   System (6.57) is said to be completely observable within n steps if and only if the state $x(j)$ can be uniquely determined from output $y(j), y(j+1), \cdots, y(j+n-1)$ for all j.

Lemma 6.1 *System (6.57) is completely reachable within n steps if and only if the rank of the reachability matrix $U_R(k)$ defined below is n $\forall k$.*

$$U_R(k) = [B(k+n-1), \ \Phi(k+n,k+n-1)B(k+n-2), \ \cdots, \tag{6.59}$$
$$\Phi(k+n,k+1)B(k)]$$

Lemma 6.2 *System (6.57) is completely observable within n steps if and only if the rank of the observability matrix defined below is n $\forall k$.*

$$U_o(k) = \begin{bmatrix} C(k) \\ C(k)\Phi(k+1,k) \\ \vdots \\ C(k+n-1)\Phi(k+n-1,k) \end{bmatrix} \tag{6.60}$$

For the engines with six cylinders where three of them (i.e., $N = 3$) share an exhaust manifold, the time-varying model (6.56) in the situation of $d = 0$ is now given by

$$\begin{cases} x(j+1) = A(j)x(j) + B(j)u(j) \\ y(j) = C(j)x(j) \end{cases} \tag{6.61}$$

with

$$A(j) = \begin{bmatrix} 0 & 0 & 0 \\ 1 & 0 & 0 \\ (1-\alpha)l_1(j) & (1-\alpha)l_2(j) & \alpha \end{bmatrix},$$

$$B(j) = \begin{bmatrix} 1 \\ 0 \\ (1-\alpha)l_3(j) \end{bmatrix}, \quad C(j) = \begin{bmatrix} 0 \\ 0 \\ 1 \end{bmatrix}^T$$

It can be confirmed that the system (6.61) is both completely reachable and observable according to Lemmas 6.1 and 6.2.

For the system (6.61), construct the following full-state observer:

$$\hat{x}(j+1) = A(k)\hat{x}(j) + B(j)u(j) + H(j)(y(j) - C(j)\hat{x}(j)) \tag{6.62}$$

where $H(j)$ is the observer gain to be designed. For the state observer construction, the goal is to find a gain matrix $H(j)$ such that the error system between the system (6.61) and the observer (6.62) is

$$e(j+1) = (A(j) - H(j)C(j))e(j) \tag{6.63}$$

where $e(j) = x(j) - \hat{x}(j)$. Under the conditions that the system (6.61) is completely reachable and observable, a procedure to find a proper $H(j)$ is presented below. In principle, the solution is achieved by using the pole placement.

Consider the following anticausal system with respect to the system (6.61):

$$\xi(j-1) = A^T(j)\xi(j) + C^T(j)v(j) \tag{6.64}$$

where $\xi \in R^n$, $v \in R^1$. By the property of the duality of the time-varying discrete system, if the pair $(A(j), C(j))$ is completely observable within n steps, then the pair $(A^T(j), C^T(j))$ is completely reachable within n steps.

According to the following fact, if the original system is observable, then there exists a time-varying vector function $\tilde{C}(j)$ such that the relative degree of the anticausal system is n between the input v and the auxiliary output σ, which is defined by

$$\sigma(j) = \tilde{C}(j)\xi(j) \tag{6.65}$$

The output matrix $\tilde{C}(j)$ can be constructed as follows.

Theorem 6.1 *If the anticausal system (6.64) is completely reachable within n steps, there exists a new output $\sigma(j)$ such that the relative degree from $v(j)$ to $\sigma(j)$ is n. And, the output matrix $\tilde{C}(j)$ can be calculated by the following equation:*

$$\tilde{C}(j) = [0, 0, 1]\bar{U}_R^{-1}(j+3) \tag{6.66}$$

where $\bar{U}_R$ denotes the anticausal reordered reachability matrix given by

$$\bar{U}_R(j) = \left[C^T(j-2), \ \Psi(j-3, j-2)C^T(j-1), \ \Psi(j-3, j-1)C^T(j) \right] \tag{6.67}$$

with the state transition matrix $\Psi(j_1, j_2)$ defined as

$$\Psi(j_1, j_2) = A^T(j_1+1) \cdots A^T(j_2-1)A^T(j_2), \quad j_1 < j_2 \tag{6.68}$$

Furthermore, for any given stable polynomials

$$q(z^{-1}) = z^{-3} + \alpha_2 z^{-2} + \alpha_1 z^{-1} + \alpha_0 \tag{6.69}$$

it is able to transform the system (6.65) into

$$q(z^{-1})\sigma(j) = H^T(j)\xi(j) + v(j) \tag{6.70}$$

where $H(j)$ is defined by

$$H^T(j) = [\alpha_0, \alpha_1, \alpha_2, 1] \begin{bmatrix} \tilde{C}(j) \\ \tilde{C}(j+1)\Psi(j+1, j) \\ \tilde{C}(j+2)\Psi(j+2, j) \\ \tilde{C}(j+3)\Psi(j+3, j) \end{bmatrix} \tag{6.71}$$

Let the state feedback input of the anticausal system (6.64) be given by

$$v(j) = -H^T(j)\xi(j) \tag{6.72}$$

Then, the pole placement state feedback control system is now obtained as

$$\xi(j-1) = (A^T(j) - C^T(j)H^T(j))\xi(j) \tag{6.73}$$

Introduce a new state variable $w(j)$, which is defined as

$$w(j) = P(j)\xi(j) \text{ with } w(j) = \begin{bmatrix} \sigma(j) \\ \sigma(j+1) \\ \sigma(j+2) \end{bmatrix} \text{ and } P(j) = \begin{bmatrix} \tilde{C}(j) \\ \tilde{C}(j+1)\Psi(j+1,j) \\ \tilde{C}(j+2)\Psi(j+2,j) \end{bmatrix} \tag{6.74}$$

Regarding the above transfer matrix $P(j)$, the following theorem is established for the development of the solution.

Theorem 6.2 *If the anticausal system (6.64) is completely reachable within three steps, the transfer matrix $P(j)$ defined in (6.74) is nonsingular for all j.*

From (6.65), (6.73), and (6.74), the following system can be deduced:

$$w(j-1) = F^*w(j) \tag{6.75}$$

where the matrix F^* is given by

$$F^* = P^T(j-1)(A^T(j) - C^T(j)H^T(j))P^{-T}(j)$$
$$= \begin{bmatrix} 0 & 1 & 0 \\ 0 & 0 & 1 \\ -\alpha_0 & -\alpha_1 & -\alpha_2 \end{bmatrix} \tag{6.76}$$

It can be noted from the obtained system (6.75) with the state transfer matrix (6.76) that the (6.74) transforms the time-varying closed-loop control system (6.73) into a time-invariant one. By Theorem 6.2, the closed-loop control system (6.73) is equivalent to the time-invariant system (6.75), which has the desired closed-loop poles, that is, $\det(z^{-1}I - F^*) = q(z^{-1})$. Finally, the full-state observer for the system (6.61) is obtained with the gain $H(j)$ given by (6.71).

6.5.3 Observer-Based Offset Learning Control

The typical structure of an iterative learning control algorithm is illustrated in Figure 6.10, where y_d denotes a given reference signal, u denotes the control input signal, and d denotes the delay steps and the memory blocks playing the role of learning.

Based on the periodic time-varying fuel path model (6.56) in [90], a learning mechanism is proposed to generate an estimation of the unknown offset $m(j)d$. For the six-cylinder engine that has three of them sharing an exhaust manifold, the $m(j)d$ periodically takes the values d_2, d_4, and d_6, which represent the offset of the individual cylinder injector (here the three cylinders of

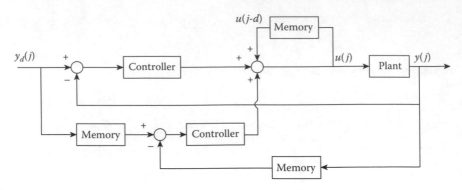

FIGURE 6.10
An illustration for the learning control principle.

the right bank, cylinders 2, 4, and 6, are taken as example). The offset can be represented by the following mathematical term:

$$m(j)d = m(j+3)d = \cdots = m(j+3n) = d_1, \quad n = 1, 2, \cdots$$

In the learning algorithm, a memory with three steps' delay is embedded in the estimation path of the offset

$$\hat{d}(j) = \hat{d}(j-3) + F(z^{-1})e(j) \tag{6.77}$$

with the learning gain function $F(z^{-1})$, which will be chosen later. Replacing the offset $m(j)d$ with this estimation in the observer (6.62), we obtain

$$\begin{cases} \hat{x}(j+1) = A(j)\hat{x}(j) + B(j)(u(j) + \hat{d}(j)) + H(j)(y(j) - C(j)\hat{x}(j)) \\ \hat{d}(j) = \hat{d}(j-3) + F(z^{-1})e(j) \end{cases} \tag{6.78}$$

The learning gain function $F(z^{-1})$ is simply chosen as the following function with respect to the current and previous learning errors:

$$F(j) = K_c + K_p z^{-1} \tag{6.79}$$

where K_c and K_p are constants.

Finally, the observer-based iterative learning control law (6.78) is embedded into the fuel injection control loop of the engine system. The structure of the control system is as shown in Figure 6.11. It can be found that in the control scheme, the offset learning algorithm provides a feedforward compensation with the estimation of the unknown offset.

6.5.4   Experimental Validation Study

To conduct the validation experiments, the parameters of the model (6.61) are identified based on the experimental data shown in Figure 6.12, which are

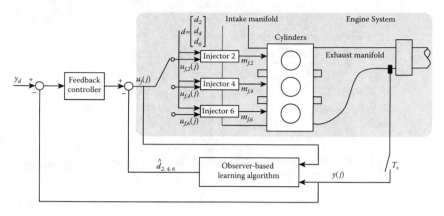

FIGURE 6.11
Observer-based learning control scheme.

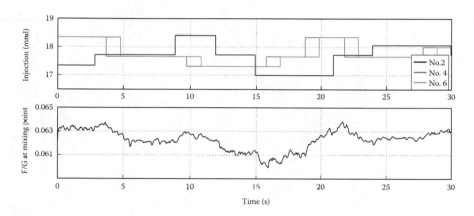

FIGURE 6.12
Experimental data for the parameter identification of model (6.61).

collected from the engine test bench. The engine operates at static mode, with the speed and load of 1,600 rpm and 80 Nm, respectively. Table 6.1 lists the identified parameter values with the RLS algorithm.

Based on the identification result shown in Table 6.3, open-loop experiments (i.e., the observer-based learning algorithm is implemented without the A/F feedback control loop) are conducted first to validate the learning control algorithm. The engine operates at the torque mode, and validation results of three cases are presented as shown in Figures 6.13 through 6.15, respectively. It can be observed from the result shown in Figure 6.13 that the output of the learning algorithm can converge to the individual offset value with respect to each cylinder. The results shown in Figures 6.14 and 6.15 are obtained with the same model parameters as in the first case. These results imply that the

TABLE 6.3
Identified parameters ($\times 10^{-3}$): engine
speed $= 1{,}600$ rpm, load $= 80$ Nm

$Mod(j,3) = r$	$r = 0$	$r = 1$	$r = 2$
$(1-\alpha)l_1(j)$	0.0767	0.0695	0.0545
$(1-\alpha)l_2(j)$	0.0428	0.0427	0.0498
α		932.229	
$(1-\alpha)l_3(j)$	0.1020	0.1534	0.1581

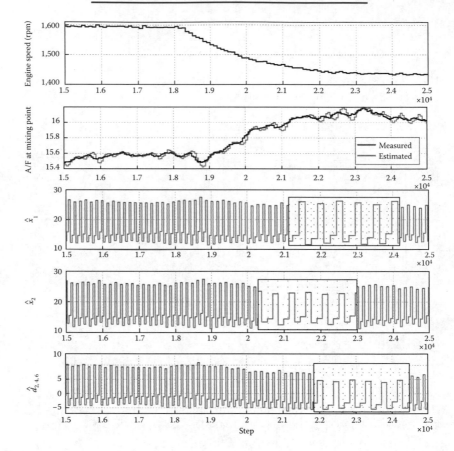

FIGURE 6.13
Experimental results with observer-based learning algorithm: load is increased
from 60 to 70 Nm during the experiment.

presented time-varying model for the fuel path of the multicylinder engine can
work well at wide operating ranges with fixed model parameters.

Finally, an experiment is conducted to validate the iterative learning control algorithm that combines with a feedback control law that is given by a
PI controller. The engine operates at the mode with the speed 1,600 rpm and

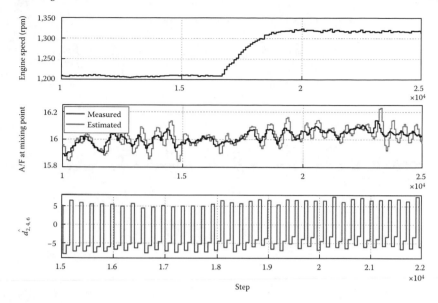

FIGURE 6.14
Experimental results with observer-based learning algorithm: load is reduced from 80 to 70 Nm during the experiment.

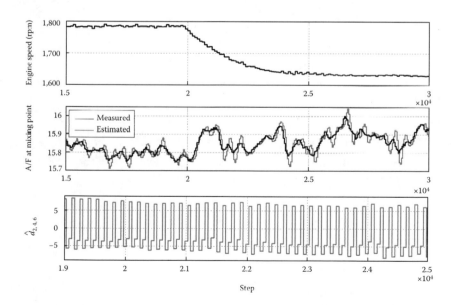

FIGURE 6.15
Experimental results with observer-based learning algorithm: load is increased from 70 to 80 Nm during the experiment.

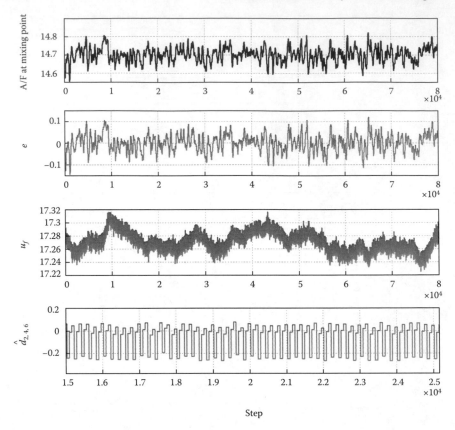

FIGURE 6.16
Experimental result with observer-based learning algorithm combined with the feedback control law.

load 60 Nm. It can be observed from the result shown in Figure 6.16 that the learning signal converges to three levels that coincide with the offset values with respect to the individual fuel path.

6.6   Conclusions

This chapter presents effective approaches for cylinder-to-cylinder A/F balancing control issues in engines with multicylinders.

Focusing on a six-cylinder engine, first, a modeling approach was proposed to realize the individual F/G estimation by using a single sensor. The individual F/G model provided an estimation of the filtered F/G for each cylinder based on the BDC-scaled multiple sampling during a cycle. The model was developed by following the exhaust gas mixing process. With the measurements of the mixing point and the values of individual cylinders as the input and the output, respectively, the linearity of the model was obtained by choosing the F/G as the control index. The presented gas mixing model directly provides the individual F/G as the model output. The identification result for the model has been demonstrated based on the experimental data.

Based on the developed individual F/G model, self-tuning control and adaptive GPC strategies were developed to deal with the A/F control issue. The experimental results with tolerant control and estimation errors show the effectiveness of the proposed scheme. However, it should be noted that the proposed individual F/G model is effective when the engine works around the corresponding static operation mode. Therefore, the control performance with the proposed control scheme also depends on the effectiveness of the individual estimation precision. In practical engineering, a technique such as gain scheduling is recommended to cope with the transient mode.

Furthermore, a periodic timing-varying model was proposed to characterize the behavior of the fuel path of the engine with multiple cylinders. On the basis of this model, the balancing control issue was also addressed via individual fuel injection and small-scale sampling and control decision. In this control scheme, attention was mainly focused on the unknown offset of each injector of the cylinders. An observer-based iterative learning algorithm to estimate the offset was proposed, and as an application of the learning algorithm, a feedback A/F control scheme with feedforward offset compensation was demonstrated. The effectiveness of the proposed learning and control algorithm can be claimed by the presented experiment results.

7

Residual Gas and Stochastic Control

7.1   Introduction

As introduced in Chapter 2, a four-stroke gasoline engine has four instinctive processes corresponding to the four strokes of each piston. For the sake of simplicity, assume that the valve-overlap angle is zero; in other words, the following induction stroke begins after the exhaust stroke ends. In induction stroke, the intake valve is open, and while the piston moves downward from top dead center (TDC) to bottom dead center (BDC), the cylinder is filled with a fresh air–fuel charge from the intake manifold for the port injection engine or with only fresh air from the intake manifold for direct injection engine, in which a fuel charge is directly injected into the cylinder in the following compression stroke. In the exhaust stroke, the exhaust valve opens, and the fluid in the combustion chamber is pushed out into the exhaust system. When the piston reaches TDC, the exhaust valve is closed and the current cycle is complete. However, the clearance volume in the cylinder is occupied by some burned gases after the exhaust stroke. The part of the burned gas remaining inside the cylinder is called residual gas. The residual gas is mixed with the fresh air–fuel charge during the intake stroke of the nest cycle.

Residual gases are sometimes also called internal exhaust gas recirculation (EGR). As is well known, the EGR technique is used to decrease the fuel consumption at partial torque output, while preserving the acceleration capabilities of the engine. Accordingly, a certain amount of residual gas is beneficial to the increase of heat capacity of the air charged. Additionally, cyclic variation of the residual gas is an important fact that affects the engine performance, such as the air–fuel (A/F) and torque generation. Thus, management of the residual gas is a considerable issue in engine performance control. However, it should be noted that it is usually not easy to obtain information on the precise amount of the residual gas mass. Consequently, research attention involves methods of the residual gas measurement besides control algorithms considering the residual gas.

One of the most popular indexes to evaluate the level of cyclic residual gas is the residual gas fraction (RGF), which is a descriptive form of the relative amount with respect to the residual gas and the total air charge in the cylinder. In the literature [91, 92], a definition of RGF is given, which is the ratio of the residual gas of the current cycle to the total gas in the next cycle. As an alternative, the RGF can also be defined as the ratio of the residual gas and the new air intake in the same cycle [93]. The intended use of RGF is that it can be estimated by using some available physical variables, such as the measured cylinder pressure and the known cylinder volume, as well as the ideal gas law and thermodynamic relationships for the engine cycle. And then, RGF can be utilized in the feedback control design for the engine performance influenced by residual gas.

To this end, a variety of results are proposed for estimating the RGF, such as in [91], [94], [95], and related references therein. Furthermore, with regard to the physics-based model, the RGF is described in [92] as a function of the engine speed, inlet pressure, exhaust pressure, valve-overlap factor, compression ratio, and A/F. The effects of these extra factors on the model of [92] are investigated in [96]. The RGF as a function of the cylinder pressure, the fresh charge temperature, and the engine speed is researched in [97].

Another difficulty in dealing with the cyclic residual gas is the stochasticity. There are a lot of factors that influence the cyclic RGF, and it is not feasible to represent the influence mathematically in general. However, it is obvious that the RGF of the current cycle is mainly determined by the combustion state of the previous cycle, and the RGF of the current cycle has a great effect on the RGF of the next cycle. Noting the stochasticity of the RGF, a natural way to describe this characteristic is to model the cyclic RGF as a Markov chain.

Meanwhile, modeling the engine dynamics taking into consideration the RGF is a further issue in addressing the control problem in the consideration of residual gas. Accordingly, based on the modeled dynamical descriptions with the RGF, control schemes are proposed for engine performance affected by the residual gas. For example, such a dynamic model is presented by [98], which introduces a random factor with Gaussian distribution into the model, describing the cyclic behavior of the A/F affected by the residual gas. Furthermore, the dynamic model is used to design a feedback control law in order to reduce the cyclic variation at lean combustion [99].

Motivated by the analysis above, in this chapter, we discuss the measurement of RGF and its statistical property and furthermore present stochastic optimal regulation and disturbance attenuation control for A/F considering the RGF. First, Section 7.2 introduces a definition of RGF, which is the ratio of the residual gas mass at the end of the exhaust stroke to the total gas mass at the end of the induction stroke of a cycle. This definition can result in deriving a cylinder pressure-based measurement of the RGF by following the physics of the inlet–exhaust process. In Section 7.3, the statistical properties of the RGF are investigated through a statistical analysis of the RGF sample. And then, a dynamical model is presented to describe the cyclic variation of the air charge, fuel charge, and combustion products under a cyclically varied RGF, where the RGF is modeled as a Markovian stochastic process [100]. In addition, stochastic regulation and disturbance rejection problems are proposed for the A/F considering the stochastic cyclic variation of the RGF. Based on the stochastic process model, an optimal controller to guarantee the accuracy of the A/F regulation [100] and a stochastic robust controller to achieve L_2 disturbance attenuation [101] are designed in Sections 7.4 and 7.5, respectively. Experiment validations are given to demonstrate the performances of the proposed control scheme in Section 7.6.

7.2 Residual Gas Measurement

In this section, for estimating the residual gas, a measurable model of the RGF will be derived. This model is obtained by utilizing the physical phenomenon of the residual gas as well as the ideal gas equation and law of thermodynamics for the engine cycle.

Consider the case for the port injection engine. The residual gas phenomenon is explained based on Figure 7.1, where k denotes the cycle-based index, BDC_e and TDC_e represent the beginning and end of the exhaust stroke, respectively, and BDC_i and TDC_i represent the end of the induction stroke and the compression stroke, respectively. Each cycle is defined from the beginning of the compression stroke to the end of the induction stroke; that is, the kth cycle is defined from $BDC_i(k)$ to $BDC_i(k+1)$. Moreover, the injection of the fresh fuel mass $M_{fn}(k)$ is completed before $BDC_i(k+1)$. Accordingly, at $BDC_i(k+1)$, the total gas mass $M_t(k+1)$ consists of the new intake gas mass $M_n(k)$ and the residual gas $M_r(k)$ at $TDC_e(k)$. $M_n(k)$ includes the fresh air mass $M_{an}(k)$ and the fresh fuel mass $M_{fn}(k)$. $M_r(k)$ is composed of the residual unburned air mass $M_{aur}(k)$, the residual unburned fuel mass $M_{fur}(k)$, and the residual combustion product mass $M_{cpr}(k)$. Similarly, $M_n(k-1)$ and $M_r(k-1)$ constitute the total gas mass $M_t(k)$ at $BDC_i(k)$. In addition, at $BDC_e(k)$, the combustion product mass $M_{cp}(k)$ contains nitric oxides (NO_x), the most of all NO, carbon monoxide (CO), sulfuric acid (H_2SO_4), particulate solid, and hydrocarbons (C_nH_m). The unburned air mass $M_{au}(k)$ includes $M_{aur}(k)$ and the discharged air during the exhaust process. The unburned fuel mass $M_{fu}(k)$ consists of $M_{fur}(k)$ and the discharged fuel during the exhaust stroke.

As shown in Figure 7.1, at $BDC_i(k)$, the total gas mass $M_t(k)$ has been inducted into the cylinder, and it is turned into $M_{au}(k)$, $M_{fu}(k)$, and $M_{cp}(k)$ in the combustion stroke. At $TDC_e(k)$, the residual gas $M_r(k)$ is trapped in the cylinder, and it is mixed with $M_n(k)$ from $TDC_e(k)$ to $BDC_i(k+1)$ to constitute $M_t(k+1)$ and begin the next cycle.

Since the amount of residual gas is often measured as a fraction of the total mass, and the definition of RGF is the ratio between the mass of residual gas and the total mass, based on the physical phenomenon of

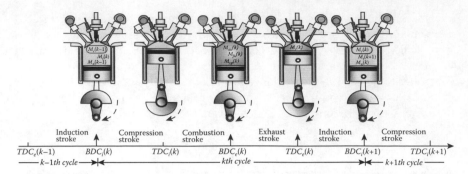

FIGURE 7.1
Residual gas phenomenon.

residual gas mentioned above, here the RGF $r(k)$ of the kth cycle is defined as follows:

$$r(k) = \frac{M_r(k)}{M_t(k)} \tag{7.1}$$

As has been shown in various papers, a feasible way to measure the RGF is an in-cylinder pressure-based method. In this section, we will pay attention to the cyclic RGF using a real-time measurement of the in-cylinder pressure at TDC_e and BDC_e. For the sake of reducing the complexity of the calculation of the RGF, the following three assumptions are made.

Assumption 7.1 The exhaust process is an adiabatic polytropic process with the polytropic constant $n = 1.3$.

Assumption 7.2 All of the gas flowing from the cylinder to the intake port during the pressure equalization process is inducted back into the cylinder.

Assumption 7.3 During the exhaust stroke, the ideal gas constant is unchanged.

To justify the value of the polytropic constant posed in Assumption 7.1, samples of the polytropic constant and the corresponding distribution parameters under the three working conditions W1, W5, and W9 (see Table 7.1) are shown in Figure 7.2, which are collected on the full-scale gasoline engine test bench described in Section 1.3. The working conditions are chosen based on the statistical properties of the RGF (see Section 7.3 for details).

In Figure 7.2, W1 (1.298, 1.56E-06) denotes the mean value μ is 1.298 and the standard deviation σ is 1.56E-06 of the sample under the operation condition W1. From Figure 7.2, it can be observed that n of each working condition is almost 1.3, with slight fluctuations.

Assumption 7.2, which can also be found in [92], keeps the residual gas mass $M_r(k)$ unchanged from the end of the exhaust stroke $TDC_e(k)$ to the

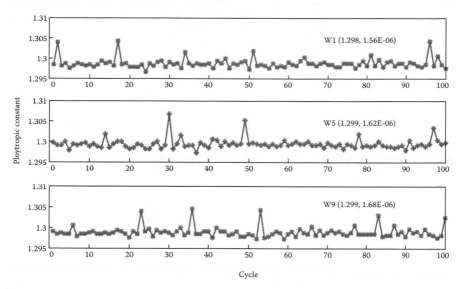

FIGURE 7.2
Polytropic constant.

end of the induction stroke $BDC_i(k+1)$. It is a foundation of the deducing process of the dynamic model with RGF.

Using the ideal gas equation, the residual gas $M_r(k)$ and the total gas mass $M_t(k)$ can be represented as follows, respectively:

$$M_r(k) = \frac{P_{TDC_e}(k)V_{TDC_e}(k)}{T_{TDC_e}(k)R_{TDC_e}(k)} \tag{7.2}$$

$$M_t(k) = \frac{P_{BDC_i}(k)V_{BDC_i}(k)}{T_{BDC_i}(k)R_{BDC_i}(k)} \tag{7.3}$$

where P, V, T, and R denote the cylinder pressure, cylinder volume, in-cylinder gas temperature, and gas constant, respectively. The subscripts TDC_e and BDC_i mean at TDC_e and BDC_i, respectively. As a result, the RGF of the kth cycle is given by

$$r(k) = \frac{M_r(k)}{M_t(k)} = \frac{P_{TDC_e}(k)V_{TDC_e}(k)T_{BDC_i}(k)R_{BDC_i}(k)}{P_{BDC_i}(k)V_{BDC_i}(k)T_{TDC_e}(k)R_{TDC_e}(k)} \tag{7.4}$$

Using the ideal gas equation again at the end of the induction stroke $BDC_i(k)$ and the beginning of the exhaust stroke $BDC_e(k)$, we have

$$\frac{P_{BDC_i}(k)V_{BDC_i}(k)}{M_{BDC_i}(k)R_{BDC_i}(k)T_{BDC_i}(k)} = \frac{P_{BDC_e}(k)V_{BDC_e}(k)}{M_{BDC_e}(k)R_{BDC_e}(k)T_{BDC_e}(k)} \tag{7.5}$$

Note that the mass of the in-cylinder gas is unchanged during the valve closed, that is,

$$M_t(k) = M_{BDC_i}(k) = M_{BDC_e}(k) \tag{7.6}$$

Thus, the relation between $T_{BDC_i}(k)$ and $T_{BDC_e}(k)$ can be derived from the observations

$$
\begin{aligned}
T_{BDC_i}(k) &= T_{BDC_e}(k) \frac{P_{BDC_i}(k)V_{BDC_i}(k)R_{BDC_e}(k)}{P_{BDC_e}(k)V_{BDC_e}(k)R_{BDC_i}(k)} \\
&= T_{BDC_e}(k) \frac{P_{BDC_i}(k)R_{BDC_e}(k)}{P_{BDC_e}(k)R_{BDC_i}(k)}
\end{aligned}
\tag{7.7}
$$

At the same time, the relation between $T_{BDC_e}(k)$ and $T_{TDC_e}(k)$ is obtained based on Assumption 7.1 as follows:

$$
T_{TDC_e}(k) = T_{BDC_e}(k) \left(\frac{P_{TDC_e}(k)}{P_{BDC_e}(k)} \right)^{\frac{n-1}{n}}
\tag{7.8}
$$

Combining (7.7) and (7.8) yields

$$
T_{TDC_e}(k) = T_{BDC_i}(k) \frac{R_{BDC_i}(k)}{R_{BDC_e}(k)} \frac{P_{BDC_e}(k)}{P_{BDC_i}(k)} \left(\frac{P_{TDC_e}(k)}{P_{BDC_e}(k)} \right)^{\frac{n-1}{n}}
\tag{7.9}
$$

Finally, by (7.4), (7.9), and Assumption 7.3, the RGF $r(k)$ can be described as

$$
r(k) = \frac{M_r(k)}{M_t(k)} = \frac{1}{\varepsilon} \left(\frac{P_{TDC_e}(k)}{P_{BDC_e}(k)} \right)^{\frac{1}{n}}
\tag{7.10}
$$

where the effective compression ratio ε is defined as the fraction of the total cylinder volume over the combustion volume.

It should be noted that Equation 7.10 is derived under the assumption that the fresh fuel mass has been injected into the cylinder at $BDC_i(k+1)$, which is the case for the port injection engine. For the direct injection gasoline engines, the fuel injection is performed during the compression stroke, but the derivation of (7.10) is still possible by only substituting $TDC_i(k)$ for $BDC_i(k)$.

7.3   Markovian Property of Residual Gas

In this section, the statistical property of the residual gas will be analyzed with the help of an experimental test for the RGF sample. The experiment is carried out on the fifth cylinder of the test bench described in Section 1.3. The resolution of the cylinder pressure sensor is set to be 15.35 pc/bar (pc is picocoulomb). Cyclic calibration and filtration are taken to improve the reliability of the cylinder pressure sensor outputs. In the cyclic calibration, the outputs of the cylinder pressure sensor are calibrated using the manifold pressure signals under the assumption that the manifold pressure and the cylinder pressure of a cycle are equivalent to each other at BDC_i. To decrease the measurement noise, such as spikes and pressure transducer random errors, a filtering algorithm is used. More specifically, the cylinder pressure at the designated point is the mean value after the removal of the maximum value and the minimum value of the multisample around the point. The samples of P_{BDC_e}, P_{TDC_e}, and RGF are exhibited in Figure 7.3.

Using this method, the RGF samples are collected under working conditions W1–W9 as given in Table 7.1. For each working condition, SA, ϕ, and M_{fn} are set to be constants, and the water temperature is 353 K.

The probability density functions and the corresponding distribution parameters are shown in Figure 7.4 and Table 7.2. From Figure 7.4, it is observed that the RGF samples obey the Gaussian distribution. According to the mean level and dispersion of the RGF samples shown in Table 7.2, W1, W5, and W9 are chosen as the typical working conditions. The autocorrelation functions of the RGF samples are given in Figure 7.5, which implies that the RGF of the current cycle has a great effect on that of the next cycle.

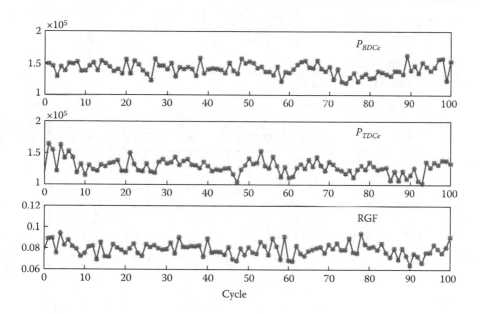

FIGURE 7.3
Samples of P_{BDC_e}, P_{TDC_e}, and RGF.

TABLE 7.1
Working conditions for RGF

Cases	W1	W2	W3	W4	W5	W6	W7	W8	W9
Engine speed (rpm)	800	1,200	1,600	800	1,200	1,600	800	1,200	1,600
Load torque (Nm)	60	60	60	90	90	90	120	120	120

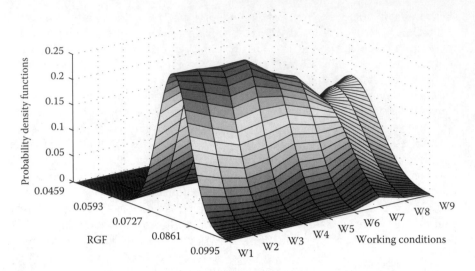

FIGURE 7.4
Probability density functions of RGF.

TABLE 7.2
Distribution parameters of the RGF samples

	W1	**W2**	**W3**	**W4**	**W5**	**W6**	**W7**	**W8**	**W9**
μ	0.0804	0.0793	0.0797	0.0796	0.0797	0.0758	0.079	0.0791	0.0724
σ	0.0056	0.0057	0.0091	0.0057	0.0066	0.0092	0.0061	0.0071	0.0097

Therefore, the RGF can be assumed to have the Markov property; that is, the cyclic RGF can be modeled as a discrete-time Markov process.

For a sample $\{X(k), k = 1, \cdots, n\}$ of a stochastic process X, the autocorrelation function r_v used is defined as

$$r_v = \frac{\sum_{k=1}^{n-v}(X(k) - \bar{X})(X(k+v) - \bar{X})}{\sum_{k=1}^{n}(X(k) - \bar{X})^2} \tag{7.11}$$

where $\bar{X}$ is the mean value of the sample and v is the lag. The autocorrelation function describes the correlation between values of the sample at different lags.

In fact, a Markovian on-line RGF model that can predict the RGF in probability during the engine test has been developed. First, by setting three thresholds, H_1, H_2, and H_3, the RGF sample is divided into four segments (see Figure 7.6). The number of sample points in each segment is the same, so that each state will have the same absence probability. Taking the mean value of each segment, the state space S of the RGF sample is obtained as

$$S = \{s_1, s_2, s_3, s_4\} \tag{7.12}$$

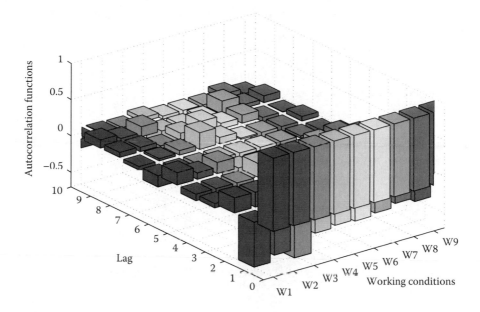

FIGURE 7.5
Autocorrelation functions of RGF.

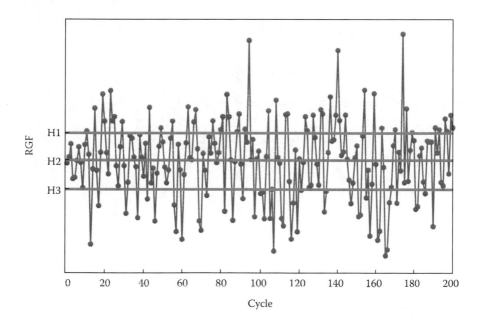

FIGURE 7.6
Thresholds of RGF sample.

where s_i is the state. The output results of off-line calculating RGF are exhibited in Figure 7.7.

And then, using the same method as in [103], the transition frequency n_{ij} can be calculated by counting the number of one-step transitions from state i to state j in Figure 7.7. By the transition frequency, the one-step transition probability matrix P is given as

$$P = \begin{pmatrix} p_{11} \ p_{12} \ p_{13} \ p_{14} \\ p_{21} \ p_{22} \ p_{23} \ p_{24} \\ p_{31} \ p_{32} \ p_{33} \ p_{34} \\ p_{41} \ p_{42} \ p_{43} \ p_{44} \end{pmatrix} \tag{7.13}$$

where

$$p_{ij} = \begin{cases} \dfrac{n_{ij}}{\sum_{j=1}^{4} n_{ij}} & \text{if } \sum_{j=1}^{4} n_{ij} > 0 \\ 0 & \text{if } \sum_{j=1}^{4} n_{ij} = 0 \end{cases} \tag{7.14}$$

Based on the state space and the one-step transition probability matrix, the one-step prediction model of the RGF at the $k+1$th cycle is

$$\hat{r}(k+1) = \sum_{j=1}^{4} s_j p_{ij} \tag{7.15}$$

where $\hat{r}(k+1)$ is the model output of the kth cycle whenever the experimentally measured value $r(k)$ of the RGF in the kth cycle is in the segment of the state s_i.

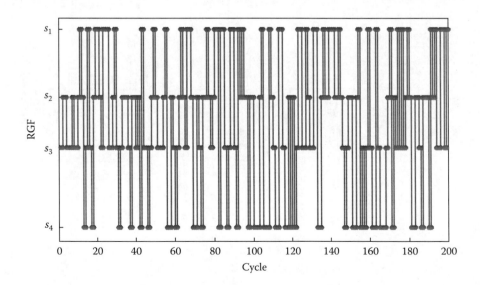

FIGURE 7.7
Results of off-line calculating RGF.

Based on the definition of the RGF given in Section 7.2, the stochastic process of an engine considering the statistical property of RGF can be obtained as follows.

With the definition of the RGF, the RGF can be rewritten as

$$r(k) = \frac{M_r(k)}{M_t(k)} = \frac{M_{aur}(k) + M_{fur}(k) + M_{cpr}(k)}{M_{an}(k-1) + M_{fn}(k-1) + M_r(k-1)} \qquad (7.16)$$

By conservation of mass, the following equality holds:

$$M_{an}(k-1) + M_{fn}(k-1) + M_r(k-1) = M_{au}(k) + M_{fu}(k) + M_{cp}(k) \qquad (7.17)$$

To simplify deduction, the fractions of residual air, residual fuel, and residual combustion products of a cycle are all assumed to be equivalent; then, utilizing (7.16) and (7.17) yields:

$$\frac{M_{aur}(k)}{M_{au}(k)} = \frac{M_{fur}(k)}{M_{fu}(k)} = \frac{M_{cpr}(k)}{M_{cp}(k)} = r(k) \qquad (7.18)$$

For the air path, the total air mass $M_a(k+1)$ at $BDC_i(k+1)$ is the amount of $M_{an}(k)$ and $M_{aur}(k)$, which is

$$M_a(k+1) = M_{an}(k) + M_{aur}(k) \qquad (7.19)$$

From (7.18), it follows that

$$M_{aur}(k) = M_{au}(k)r(k) \qquad (7.20)$$

The value of $M_{au}(k)$ is equal to $M_a(k)$ minus the air mass consumed in the combustion process $M_f(k)\eta_c(k)\lambda_d$:

$$M_{au}(k) = M_a(k) - M_f(k)\eta_c(k)\lambda_d \qquad (7.21)$$

where λ_d denotes the theoretical A/F. $\eta_c(k)$ is the combustion efficiency, and its normal definition is given as the following form by [104]:

$$\eta_c(k) = \frac{Q(k)}{M_{fn}(k-1)H_u} \qquad (7.22)$$

which means the fraction of the net chemical energy release of fuel due to combustion Q over the fresh fuel's chemical energy, where H_u is the fuel low heating value. Taking into consideration the residual fuel, (7.22) is rewritten as

$$\eta_c(k) = \frac{Q(k)}{(M_{fn}(k-1) + M_{fur}(k-1))H_u} = \frac{Q(k)}{M_f(k)H_u} \qquad (7.23)$$

while $Q(k)$ can be represented as

$$Q(k) = (M_f(k) - M_{fu}(k))H_u \qquad (7.24)$$

Substituting (7.24) into (7.23), η_c is defined as the fraction of the burned fuel mass over the total fuel mass:

$$\eta_c(k) = \frac{M_f(k) - M_{fu}(k)}{M_f(k)} \tag{7.25}$$

Combining (7.19), (7.20), and (7.21), the dynamics with RGF of the air path is

$$M_a(k+1) = (M_a(k) - M_f(k)\eta_c(k)\lambda_d)r(k) + M_{an}(k) \tag{7.26}$$

For the fuel path, the total fuel mass $M_f(k+1)$ at $BDC_i(k+1)$ is the sum of $M_{fn}(k)$ and $M_{fur}(k)$:

$$M_f(k+1) = M_{fn}(k) + M_{fur}(k) \tag{7.27}$$

From (7.18), the relation between $M_{fur}(k)$ and $M_{fu}(k)$ is

$$M_{fur}(k) = M_{fu}(k)r(k) \tag{7.28}$$

By the definition of $\eta_c(k)$ (7.25), the following equality holds:

$$M_{fu}(k) = M_f(k)(1 - \eta_c(k)) \tag{7.29}$$

From (7.27), (7.28), and (7.29), the dynamics with RGF of the fuel path is

$$M_f(k+1) = M_f(k)(1 - \eta_c(k))r(k) + M_{fn}(k) \tag{7.30}$$

For the combustion products' path, the combustion products' mass $M_{cp}(k+1)$ is equal to $M_{cpr}(k)$ plus $M_{cpn}(k+1)$ generated by the combustion process of the $k+1$th cycle:

$$M_{cp}(k+1) = M_{cpr}(k) + M_{cpn}(k+1) \tag{7.31}$$

By conservation of mass, $M_{cpn}(k+1)$ is given by

$$\begin{aligned} M_{cpn}(k+1) &= \lambda_d M_f(k+1)\eta_c(k+1) + M_f(k+1)\eta_c(k+1) \\ &= (\lambda_d + 1)M_f(k+1)\eta_c(k+1) \end{aligned} \tag{7.32}$$

By substituting (7.32) into (7.31), the dynamics with RGF of the combustion products is obtained as follows.

$$M_{cp}(k+1) = M_{cpr}(k) + (\lambda_d + 1)M_f(k+1)\eta_c(k+1) \tag{7.33}$$

Based on (7.26), (7.30), and (7.33), the dynamic model with RGF is expressed as follows:

$$\begin{cases} M_a(k+1) = (M_a(k) - \lambda_d\eta_c(k)M_f(k))r(k) + M_{an}(k) \\ M_f(k+1) = M_f(k)(1 - \eta_c(k))r(k) + M_{fn}(k) \\ M_{cp}(k+1) = M_{cpr}(k) + (\lambda_d + 1)M_f(k+1)\eta_c(k+1) \end{cases} \tag{7.34}$$

Similar to Equation 7.10, (7.34) is derived under the assumption that $M_{fn}(k)$ has been injected into the cylinder at $BDC_i(k+1)$, which is the case for the port injection engine. For the direct injection engines, the fuel injection is performed during the compression stroke, and the derivation of (7.34) is still possible by substituting $TDC_i(k)$ for $BDC_i(k)$.

7.4 Stochastic Optimal Regulation

As is well known, the A/F of a cycle is determined by the air mass, the fuel mass, and the residual gas mass. Hence, taking the cyclic RGF into account and utilizing the stochastic process of the air path and fuel path modeled in Section 7.3, a stochastic optimal A/F regulation problem is formulated in this section. Define

$$y(k) = M_a(k) - M_f(k)\lambda_d \qquad (7.35)$$

Then, considering the dynamic equations of the air and fuel paths of (7.34), and the fresh air mass consisting of the nominal fresh air mass M_{an0} and the air perturbation ΔM_{an},

$$M_{an}(k) = M_{an0}(k) + \Delta M_{an}(k) \qquad (7.36)$$

and assuming that the known nominal fresh fuel mass $M_{fn0} = \frac{M_{an0}}{\lambda_d}$, the following dynamical equation of the A/F results:

$$y(k+1) = r(k)y(k) + u(k) + \Delta M_{an}(k) \qquad (7.37)$$

where

$$u(k) = M_{an0} - M_{fn}(k)\lambda_d = (M_{fn0} - M_{fn}(k))\lambda_d \qquad (7.38)$$

For the sake of simplicity, only the known nominal fresh air mass M_{an0} is considered in the controller design process, that is,

$$y(k+1) = r(k)y(k) + u(k) \qquad (7.39)$$

where $y(k)$ is regarded as the system state, $u(k)$ is the control input signal, and the system matrix $A(r(k)) = r(k)$, $r(k)$ is a discrete-time, discrete-state Markov process with the one-step transition probability of the RGF p_{ij}, that is, $P\{r(k+1) = s_j | r(k) = s_i\} = p_{ij}$ $(i, j \in \{1, \cdots, s\})$ (see Section 7.3). The cost criterion is given by

$$J(u) = E\left\{\sum_{k=0}^{N-1}\left(c_1 y^2(k) + c_2\left(\frac{u(k)}{\lambda_d}\right)^2\right)\right\} \qquad (7.40)$$

where $c_1 > 0$ and $c_2 > 0$ are the tuning parameters for the desired closed-loop performance and E is the expectation operator taken with respect to the RGF.

The control input utilizes feedback of both the state and the operating mode $u(k) = \bar{u}(k, y(k), s_i)$, the control policy $\bar{u}$ must be selected from an admissible set $U = R$. Hence, the optimal control problem is formulated as follows: to find an optimal controller $u^* \in U$ to minimize the cost function (7.40), that is,

$$
\begin{cases}
V(k, y, s_i) = J(u^*) \\
\qquad = \min_{\bar{u} \in U} E \left\{ \sum_{l=k}^{N-1} \left(c_1 y^2(l) + c_2 \left(\frac{\bar{u}(k, y(l), r(l))}{\lambda_d} \right)^2 \right) \middle| y(k) = y, r(k) = s_i \right\} \\
\qquad\qquad\qquad\qquad\qquad\qquad\qquad\qquad\qquad k = 0, 1, \cdots, N-1 \\
V(N, y(N), r(N)) = 0
\end{cases}
$$

$$(7.41)$$

An explicit expression for u^* will be found by applying the formalism of dynamic programming to (7.41). First, using the optimality principle, (7.41) is written as the form

$$
\begin{cases}
V(k, y, s_i) = \min_{\bar{u} \in U} E \left\{ \left(c_1 y^2(k) + c_2 \left(\frac{\bar{u}(k)}{\lambda_d} \right)^2 \right) \right. \\
\qquad\qquad\qquad \left. + V(k+1, y(k+1), r(k+1) = s_j) \middle| k, r(k) = s_i \right\} \\
\qquad\qquad\qquad\qquad\qquad\qquad\qquad\qquad k = 0, 1, \cdots, N-1 \\
V(N, y(N), r(N)) = 0
\end{cases}
$$

$$(7.42)$$

As a trial solution to (7.42), let the value function be constructed as

$$V(k, y, s_i) = c_3(k, s_i) y^2(k) \tag{7.43}$$

By substituting (7.43) and (7.39) into (7.42), we obtain

$$
c_3(k, s_i) y^2(k) = \min_{\bar{u} \in U} \left\{ \left(c_1 y^2(k) + c_2 \left(\bar{u}(k)/\lambda_d \right)^2 \right) \right.
$$
$$
\left. + (s_i y(k) + \bar{u}(k))^2 E[c_3(k+1, r(k+1) = s_j) | k, r(k) = s_i] \right\}
$$

$$(7.44)$$

Since

$$E[c_3(k+1, r(k+1) = s_j) | k, r(k) = s_i] = \sum_{j=1}^{s} c_3(k+1, s_j) p_{ij}$$

then

$$
c_3(k, s_i) y^2(k) = \min_{\bar{u} \in U} \left\{ \left(c_1 y^2(k) + c_2 \left(\bar{u}(k)/\lambda_d \right)^2 \right) \right.
$$
$$
\left. + (s_i y(k) + \bar{u}(k))^2 \sum_{j=1}^{s} c_3(k+1, s_j) p_{ij} \right\}
$$

$$(7.45)$$

Differentiating (7.45) with respect to $\bar{u}(k)$ yields the optimal control:

$$u^*(k, y(k), s_i) = -\frac{s_i \sum_{j=1}^{s} c_3(k+1, s_j)p_{ij}}{c_2/\lambda_d^2 + \sum_{j=1}^{s} c_3(k+1, s_j)p_{ij}} y(k) \qquad (7.46)$$

Substituting (7.46) into (7.45) produces

$$c_3(k, s_i) = c_1 + s_i^2 \sum_{j=1}^{s} c_3(k+1, s_j)p_{ij}$$

$$-\frac{\left(s_i \sum_{j=1}^{s} c_3(k+1, s_j)p_{ij}\right)^2}{c_2/\lambda_d^2 + \sum_{j=1}^{s} c_3(k+1, s_j)p_{ij}}, \qquad k = 0, 1, \cdots, N-1 \qquad (7.47)$$

$$c_3(N, r(N)) = 0$$

Using (7.38) and (7.46) yields

$$M_{fn}^*(k, y(k), s_i) = (M_{an0} - u^*(k, s_i))/\lambda_d$$

$$= \frac{1}{\lambda_d}\left(M_{an0} + \frac{s_i \sum_{j=1}^{s} c_3(k+1, s_j)p_{ij}y(k)}{c_2/\lambda_d^2 + \sum_{j=1}^{s} c_3(k+1, s_j)p_{ij}}\right) \qquad (7.48)$$

The optimal controller of the infinite optimization interval can be obtained by iterating (7.48) and (7.47) to the steady value

$$M_{fn}^*(\infty, y(k), s_i) = \frac{1}{\lambda_d}\left(M_{an0} + \frac{s_i \sum_{j=1}^{s} c_3(\infty, s_j)p_{ij}y(k)}{c_2/\lambda_d^2 + \sum_{j=1}^{s} c_3(\infty, s_j)p_{ij}}\right) \qquad (7.49)$$

and

$$c_3(\infty, s_i) = c_1 + s_i^2 \sum_{j=1}^{s} c_3(\infty, s_j)p_{ij} - \frac{\left(s_i \sum_{j=1}^{s} c_3(\infty, s_j)p_{ij}\right)^2}{c_2/\lambda_d^2 + \sum_{j=1}^{s} c_3(\infty, s_j)p_{ij}} \qquad (7.50)$$

When the fresh air mass perturbation ΔM_{an} is not neglected, which is regarded as a Gaussian white noise with $E(\Delta M_{an}(k)) = 0$ and $E(\Delta M_{an}^2(k)) = d$, the closed-loop system consisting of the dynamic equations of the air and fuel paths

$$\begin{cases} M_a(k+1) = r(k)(M_a(k) - \lambda_d M_f(k)\eta_c(k)) + M_{an}(k) \\ M_f(k+1) = r(k)(1 - \eta_c(k))M_f(k) + M_{fn}(k) \end{cases} \qquad (7.51)$$

and (7.49) with (7.50) is mean square stable for any initial condition $(M_a(0), M_f(0))^T$, initial distribution $r(0)$, and disturbance $\Delta M_{an}(k)$.

This result can be proved by the following lemma from [102].

Lemma 7.1 *For the system*

$$x(k+1) = A(r(k))x(k) + B(r(k))w(k) \qquad (7.52)$$

with the state $x(k)$, Gaussian white noise $w(k)$, and finite state Markov chain $r(k)$ taking values in $S = \{s_1, s_2, \cdots, s_s\}$, if for any given set of symmetric

matrices $\{W(s_i) > 0, i = 1, \cdots, s\}$ there exists a set of symmetric solutions
$\{\chi(s_i) > 0, i = 1, \cdots, s\}$ such that

$$\sum_{j=1}^{s} p_{ij} A^T(s_i) \chi(s_j) A(s_i) - \chi(s_i) = -W(s_i) \qquad (7.53)$$

then the system (7.52) is mean square stable.

Substituting (7.49) and (7.50) into (7.51), the following form can be obtained:

$$\begin{pmatrix} M_a(k+1) \\ M_f(k+1) \end{pmatrix} = \begin{pmatrix} A_{11} & A_{12} \\ A_{21} & A_{22} \end{pmatrix} \begin{pmatrix} M_a(k) \\ M_f(k) \end{pmatrix} + \begin{pmatrix} M_{an0} \\ M_{an0}/\lambda_d \end{pmatrix} + \begin{pmatrix} \Delta M_{an}(k) \\ 0 \end{pmatrix}$$
$$(7.54)$$

where

$$A_{11} = r(k), \quad A_{12} = -\lambda_d r(k) \eta_c(k)$$

$$A_{21} = \frac{r(k) \sum_{j=1}^{s} c_3(\infty, s_j) p_{ij}}{\lambda_d (c_2/\lambda_d^2 + \sum_{j=1}^{s} c_3(\infty, s_j) p_{ij})}$$

$$A_{22} = \frac{r(k)(1 - \eta_c(k)) c_2/\lambda_d^2 - r(k) \eta_c(k) \sum_{j=1}^{s} c_3(\infty, s_j) p_{ij}}{c_2/\lambda_d^2 + \sum_{j=1}^{s} c_3(\infty, s_j) p_{ij}}$$

Clearly, (7.54) meets the form of (7.52) with $A(r(k)) = \begin{pmatrix} A_{11} & A_{12} \\ A_{21} & A_{22} \end{pmatrix}$, and
choosing

$$\chi(s_i) = \chi = \begin{pmatrix} 1/\lambda_d & 0 \\ 0 & \lambda_d \end{pmatrix}, \quad i = 1, \cdots, s \qquad (7.55)$$

the following equality holds:

$$\sum_{j=1}^{s} p_{ij} A^T(s_i) \chi A(s_i) - \chi = -W \qquad (7.56)$$

with

$$W = \begin{pmatrix} 1/\lambda_d - A_{11}^2/\lambda_d - \lambda_d A_{21}^2 & -A_{11} A_{12}/\lambda_d - \lambda_d A_{21} A_{22} \\ -A_{11} A_{12}/\lambda_d - \lambda_d A_{21} A_{22} & \lambda_d - A_{12}^2/\lambda_d - \lambda_d A_{22}^2 \end{pmatrix}$$

Noting that the RGF $r(k) \in (0, 0.5)$ and $\eta_c(k) \in (0, 1)$, it follows that

$$\det(1/\lambda_d - A_{11}^2/\lambda_d - \lambda_d A_{21}^2) > 0, \quad \det(W) > 0$$

That is, W is a symmetric positive-definite matrix. By Lemma 7.1, the closed-loop system is mean square stable.

7.5   Disturbance Rejection

In this section, based on the discrete-time dynamic model (7.37) of the A/F, where the cycle-by-cycle fluctuation of the fresh air $\Delta M_{an}(k)$ is assumed to be a disturbance belonging to $\mathcal{L}_2[0, \infty]$ (the space of square summable vector sequence over $[0, \infty)$), a stochastic robust controller is designed to achieve L_2 disturbance attenuation, that is, for a given scalar $\gamma > 0$, to find a feedback controller

$$M_{fn}(k) = \frac{M_{an0}}{\lambda_d} + \alpha(r(k))y(k) \tag{7.57}$$

such that the inequality

$$E \sum_{k=0}^{\infty} y^2(k) < \gamma \sum_{k=0}^{\infty} \Delta M_{an}^2(k) \tag{7.58}$$

holds for all $\Delta M_{an}(k)$. Furthermore, with the feedback controller (7.57), the system (7.51) of the air path and fuel paths is mean square stable.

Define the fuel cyclic variation $\Delta M_{fn}(k)$ as

$$\Delta M_{fn}(k) = M_{fn}(k) - M_{fn0} \tag{7.59}$$

where M_{fn0} is the known nominal fuel mass that can be obtained as $M_{fn0} = M_{an0}/\lambda_d$. Accordingly, the dynamic equation (7.37) is rewritten as

$$y(k+1) = r(k)y(k) - \lambda_d \Delta M_{fn}(k) + \Delta M_{an}(k) \tag{7.60}$$

where $r(k)$ is assumed to be a discrete-time Markov chain with the one-step transition probability of the RGF p_{ij} (see Section 7.3).

Therefore, the following proposition can be concluded.

Proposition 7.1   *For the system (7.60), given a scalar $\gamma > 0$, if there exists a set $\{\chi(s_i) > 0, s_i \in S\}$ such that the following inequalities hold,*

$$\gamma - \sum_{j=1}^{s} p_{ij} \chi(s_j) > 0, \quad R(s_i) < 0 \tag{7.61}$$

with

$$R(s_i) = \sum_{j=1}^{s} p_{ij} \chi(s_j)s_i^2 + \frac{2(\sum_{j=1}^{s} p_{ij} \chi(s_j)s_i)^2}{\gamma - \sum_{j=1}^{s} p_{ij} \chi(s_j)} + 1 - \chi(s_i) \\ - \frac{(\gamma - \sum_{j=1}^{s} p_{ij} \chi(s_j)) \sum_{j=1}^{s} p_{ij} \chi(s_j)s_i^2}{\gamma + \sum_{j=1}^{s} p_{ij} \chi(s_j)} \tag{7.62}$$

the fuel variation

$$\Delta M_{fn}(k) = \alpha(r(k))y(k) \tag{7.63}$$

with

$$\alpha(r(k) = s_i) = \frac{s_i(\gamma - \sum_{j=1}^{s} p_{ij} \chi(s_j))}{\lambda_d(\gamma + \sum_{j=1}^{s} p_{ij} \chi(s_j))} \tag{7.64}$$

can guarantee the system (7.60) has an L_2-gain performance γ over $[0, \infty)$.
Moreover, with the feedback controllers (7.57), (7.63), and (7.64), the system
(7.51) of the air path and fuel paths is mean square stable.

Proof. Choose a stochastic Lyapunov function as

$$V(k, r(k) = s_i) = \chi(s_i)y^2(k) \tag{7.65}$$

and consider (7.60) and (7.65); the following equality holds:

$$\begin{aligned}
\Delta V(k, r(k) = s_i) &= E[V(k+1, r(k+1))|r(k) = s_i] - V(k, r(k) = s_i) \\
&= \sum_{j=1}^{s} p_{ij} \chi(s_j)(s_i y(k) - \lambda_d \Delta M_{fn}(k) + \Delta M_{an}(k))^2 \\
&\quad - \chi(s_i)y^2(k)
\end{aligned} \tag{7.66}$$

By completing the square, (7.66) satisfies the following inequality:

$$\begin{aligned}
\Delta V(k, r(k) = s_i) &\leq R(s_i)y^2(k) - y^2(k) + \gamma \Delta M_{an}^2(k) \\
&\quad + (A(s_i)y(k) - B(s_i)\Delta M_{fn}(k))^2
\end{aligned} \tag{7.67}$$

where

$$A(s_i) = \frac{[\sum_{j=1}^{s} p_{ij} \chi(s_j)(\gamma - \sum_{j=1}^{s} p_{ij} \chi(s_j))]^{\frac{1}{2}} s_i}{[\gamma + \sum_{j=1}^{s} p_{ij} \chi(s_j)]^{\frac{1}{2}}} \tag{7.68}$$

$$B(s_i) = \frac{[\sum_{j=1}^{s} p_{ij} \chi(s_j)(\gamma + \sum_{j=1}^{s} p_{ij} \chi(s_j))]^{\frac{1}{2}} \lambda_d}{[\gamma - \sum_{j=1}^{s} p_{ij} \chi(s_j)]^{\frac{1}{2}}} \tag{7.69}$$

By (7.61) through (7.64), it follows that

$$\Delta V(k, r(k) = s_i) \leq -y^2(k) + \gamma \Delta M_{an}^2(k) \tag{7.70}$$

This means that

$$E[V(k+1, r(k+1))|r(k) = s_i] - V(k, r(k) = s_i) \leq -y^2(k) + \gamma \Delta M_{an}^2(k) \tag{7.71}$$

For every s_i, (7.71) is true; therefore,

$$\begin{aligned}
EV(k+1, &r(k+1)) - EV(k, r(k)) \\
&= \sum_{i=1}^{s} (E[V(k+1, r(k+1))|r(k) = s_i] - V(k, s_i))p_{s_i} \\
&\leq \sum_{i=1}^{s} (-y^2(k) + \gamma \Delta M_{an}^2(k))p_{s_i} \\
&= -Ey^2(k) + \gamma \Delta M_{an}^2(k)
\end{aligned} \tag{7.72}$$

By summing both sides from 0 to ∞, the following inequality holds:

$$EV(\infty, r(\infty)) - EV(0, r(0)) \leq -E\sum_{k=0}^{\infty} y^2(k) + \gamma \sum_{k=0}^{\infty} \Delta M_{an}^2(k) \qquad (7.73)$$

Hence,

$$E\sum_{k=0}^{\infty} y^2(k) \leq V(0, r(0)) + \gamma \sum_{k=0}^{\infty} \Delta M_{an}^2(k) \qquad (7.74)$$

Now considering the system consisting of (7.51) and (7.63) with (7.64)

$$\begin{pmatrix} M_a(k+1) \\ M_f(k+1) \end{pmatrix} = \begin{pmatrix} A_{11} & A_{12} \\ A_{21} & A_{22} \end{pmatrix} \begin{pmatrix} M_a(k) \\ M_f(k) \end{pmatrix} + \begin{pmatrix} M_{an0} \\ M_{an0}/\lambda_d \end{pmatrix} + \begin{pmatrix} \Delta M_{an}(k) \\ 0 \end{pmatrix} \qquad (7.75)$$

where

$$A_{11} = r(k), \quad A_{12} = -\lambda_d r(k)\eta_c(k)$$

$$A_{21} = \frac{s_i(\gamma - \sum_{j=1}^{s} p_{ij}\chi(s_j))}{\lambda_d(\gamma + \sum_{j=1}^{s} p_{ij}\chi(s_j))}$$

$$A_{22} = (1 - \eta_c(k))r(k) - \frac{s_i(\gamma - \sum_{j=1}^{s} p_{ij}\chi(s_j))}{\gamma + \sum_{j=1}^{s} p_{ij}\chi(s_j)}$$

and choosing

$$\chi(s_i) = \chi = \begin{pmatrix} 1/\lambda_d & 0 \\ 0 & \lambda_d \end{pmatrix}, \quad i = 1, \cdots, s \qquad (7.76)$$

it follows that

$$\sum_{j=1}^{N} p_{ij} A_{s_i}^T \chi A_{s_i} - \chi = -W \qquad (7.77)$$

where A_{s_i} means $r(k) = s_i$ in matrix $A = \begin{pmatrix} A_{11} & A_{12} \\ A_{21} & A_{22} \end{pmatrix}$ and

$$W = \begin{pmatrix} 1/\lambda_d - A_{11}^2/\lambda_d - \lambda_d A_{21}^2 & -A_{11}A_{12}/\lambda_d - \lambda_d A_{21}A_{22} \\ -A_{11}A_{12}/\lambda_d - \lambda_d A_{21}A_{22} & \lambda_d - A_{12}^2/\lambda_d - \lambda_d A_{22}^2 \end{pmatrix} \qquad (7.78)$$

By noting that the RGF $r(k) \in (0, 0.5)$ and $\mu \in (0, 1)$, the following inequalities hold:

$$\det(1/\lambda_d - A_{11}^2/\lambda_d - \lambda_d A_{21}^2) > 0, \quad \det(W) > 0 \qquad (7.79)$$

That is, W is a symmetric positive-definite matrix. Hence, the system (7.75) is mean square stable.

7.6 Experimental Case Studies

In this section, the proposed control schemes in Sections 7.4 and 7.5 are validated in the test benchmark of the engine described in Section 1.3.

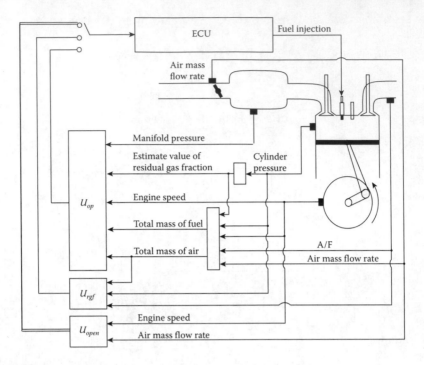

FIGURE 7.8
Optimal control scheme for A/F regulation.

First, the optimal controller designed in Section 7.4 is validated. Without loss of generality, the experimental tests of W2 and W6 are performed to verify the proposed control strategy, and the control scheme is shown in Figure 7.8, where U_{op} denotes the controller designed in Section 7.4, and U_{rgf} and U_{open} are given by the following:

$$U_{rgf}(k) = \frac{M_{an}(k)}{\lambda_d} + \frac{M_{aur}(k)}{\lambda_d} - M_{fur}(k)$$

$$U_{open}(k) = \frac{M_{an}(k)}{\lambda_d}$$

If the M_{an0} values of W2 and W6 are 0.186 and 0.231 g, respectively, then the corresponding M_{fn0} values calculated by $M_{fn0} = M_{an0}/\lambda_d$ are 0.0128 and 0.0159 g. The state spaces and one-step transition probability matrices of the RGF of W2 and W6 are given in (7.80) and (7.81).

$$S_{W2} = \{0.085,\ 0.082,\ 0.080,\ 0.076\},\quad S_{W6} = \{0.087,\ 0.081,\ 0.075,\ 0.072\} \tag{7.80}$$

$$P_{W2} = \begin{pmatrix} 0.10 & 0.33 & 0.39 & 0.27 \\ 0.11 & 0.14 & 0.39 & 0.36 \\ 0.31 & 0.14 & 0.14 & 0.41 \\ 0.41 & 0.31 & 0.10 & 0.18 \end{pmatrix},\quad P_{W6} = \begin{pmatrix} 0.25 & 0.02 & 0.19 & 0.54 \\ 0.51 & 0.24 & 0.04 & 0.21 \\ 0.21 & 0.52 & 0.23 & 0.04 \\ 0.03 & 0.21 & 0.55 & 0.21 \end{pmatrix} \tag{7.81}$$

Setting $c_1 = 1$ and $c_2 = 1$, we can obtain

$$c_{3,\text{W2}}(\infty) = \begin{cases} 1.008 & \text{if } r(k) = 0.085 \\ 1.0075 & \text{if } r(k) = 0.082 \\ 1.0072 & \text{if } r(k) = 0.080 \\ 1.0065 & \text{if } r(k) = 0.076 \end{cases}$$

$$(7.82)$$

$$c_{3,\text{W6}}(\infty) = \begin{cases} 1.0084 & \text{if } r(k) = 0.087 \\ 1.0073 & \text{if } r(k) = 0.081 \\ 1.0063 & \text{if } r(k) = 0.075 \\ 1.0059 & \text{if } r(k) = 0.072 \end{cases}$$

Samples of the A/F and the fresh fuel mass of W2 and W6 are exhibited in Figures 7.9 and 7.10.

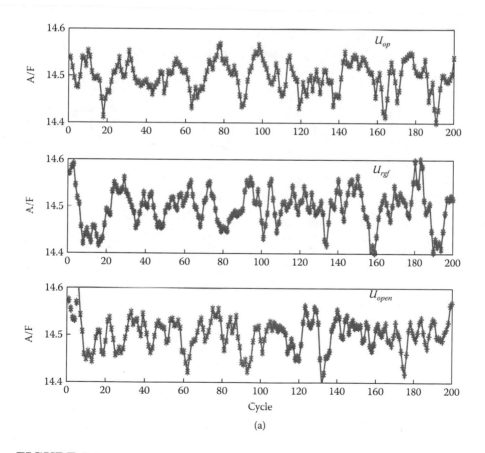

(a)

FIGURE 7.9

Control performance of W2. (a) A/F of W2. *(Continued)*

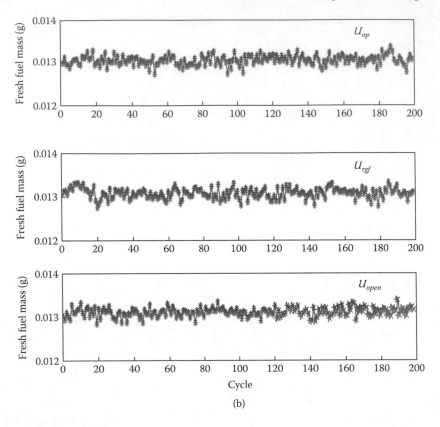

(b)

FIGURE 7.9 (Continued)
Control performance of W2. (b) Fresh fuel mass of W2.

The performance indexes of W2 are listed in Table 7.3, and those of W6 in Table 7.4.

It can be seen from Table 7.3 that J of U_{op} is smaller than those of U_{rgf} and U_{open}. Therefore, the control performance of U_{op} is better than those of U_{rgf} and U_{open}. Different from the simulations, J of U_{rgf} is larger than the one of U_{open}. This different trend is caused by the measurement noises of the additional sensors that U_{rgf} requires, such as cylinder pressure sensor and A/F sensor. Similar results can be seen in Table 7.4.

In the following, the proposed control scheme in Section 7.5 will be validated.

The experiments have two working conditions $\mathcal{A}$ and $\mathcal{B}$, which is listed in Table 7.5.

The RGF measurement is obtained by (7.10) described in Section 7.2.

The RGF samples of $\mathcal{A}$ and $\mathcal{B}$ are shown in Figure 7.11.

The RGF probability density functions and the autocorrelation functions of $\mathcal{A}$ and $\mathcal{B}$ are shown in Figure 7.12.

By setting two thresholds, the RGF samples of $\mathcal{A}$ and $\mathcal{B}$ are divided into three segments:

$$r_{\mathcal{A}}(k) = \begin{cases} 0.0846, & \text{if } r(k) \in (0.0817, \infty) \\ 0.0798, & \text{if } r(k) \in (0.0781, 0.0817) \\ 0.0755, & \text{if } r(k) \in (-\infty, 0.0781) \end{cases}$$

$$r_{\mathcal{B}}(k) = \begin{cases} 0.0871, & \text{if } r(k) \in (0.083, \infty) \\ 0.0791, & \text{if } r(k) \in (0.0753, 0.083) \\ 0.0699, & \text{if } r(k) \in (-\infty, 0.0753) \end{cases}$$

Taking the mean value of each segment, the state spaces of $\mathcal{A}$ and $\mathcal{B}$ are given by

$$S_{\mathcal{A}} = \{s_1, s_2, s_3\} = \{0.0846, 0.0798, 0.0755\},$$
$$S_{\mathcal{B}} = \{s_1, s_2, s_3\} = \{0.0871, 0.0791, 0.0699\}$$

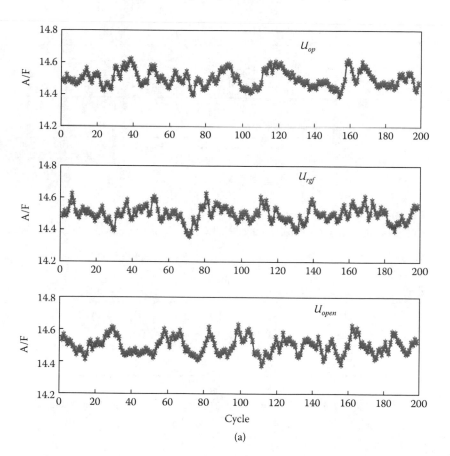

FIGURE 7.10

Control performance of W6. (a) A/F of W6. *(Continued)*

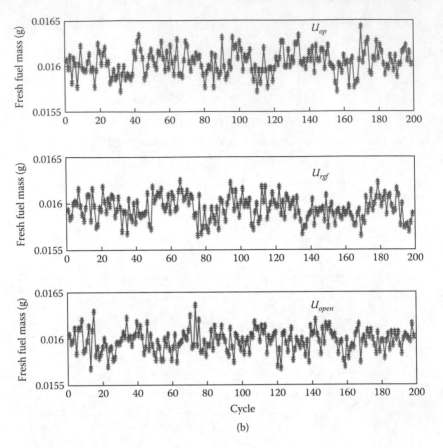

FIGURE 7.10 (Continued)
Control performance of W6. (b) Fresh fuel mass of W6.

TABLE 7.3
Performance indexes of W2

	J	$\|\text{mean}(\lambda - \lambda_d)\|/\lambda_d$	$\text{var}(\lambda - \lambda_d)$	J_u
U_{op}	3.56E − 05	2.46E − 05	1.0E − 03	2.63E − 06
U_{rgf}	4.30E − 05	6.64E − 05	1.2E − 03	3.48E − 06
U_{open}	4.23E − 05	3.01E − 05	1.3E − 03	2.49E − 06

By counting the number of the one-step transition frequencies of the state, the one-step transition probability matrices are obtained as follows:

$$
P_{\mathcal{A}} = \begin{pmatrix} 0.42 & 0.28 & 0.3 \\ 0.37 & 0.35 & 0.28 \\ 0.22 & 0.36 & 0.42 \end{pmatrix}, \quad P_{\mathcal{B}} = \begin{pmatrix} 0.29 & 0.46 & 0.25 \\ 0.18 & 0.31 & 0.51 \\ 0.53 & 0.22 & 0.25 \end{pmatrix}
$$

TABLE 7.4

Performance indexes of W6

	J	$|\text{mean}(\lambda - \lambda_d)|/\lambda_d$	$\text{var}(\lambda - \lambda_d)$	J_u
U_{op}	$3.09E - 04$	$3.17E - 05$	$2.3E - 03$	$1.01E - 05$
U_{rgf}	$3.63E - 04$	$3.07E - 05$	$3.7E - 03$	$1.36E - 05$
U_{open}	$3.52E - 04$	$4.72E - 05$	$2.6E - 03$	$9.26E - 06$

TABLE 7.5

Working conditions for disturbance attenuation

Cases	External Load	Engine Speed	Water Temperature	Throttle Angle
$\mathcal{A}$	50 Nm	1,000 rpm	80°C	4.5°
$\mathcal{B}$	70 Nm	1,400 rpm	80°C	7.2°

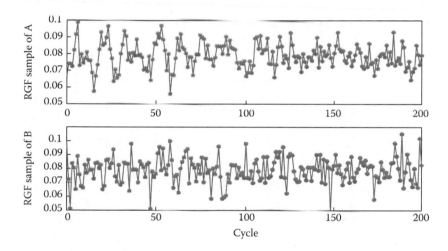

FIGURE 7.11

RGF samples.

The experiment tests of $\mathcal{A}$ and $\mathcal{B}$ are performed to verify the proposed control strategy. The scalar γ is set to be 1.4 and χ is chosen as 1.2.

Figure 7.13 shows the performance of the A/F of the proposed controller. Meanwhile, the comparisons with the open-loop controller $U_{open} = M_{an}(k)/\lambda_d$ in the two cases are also given in Figure 7.13.

The performances of the total air mass and the total fuel mass of $\mathcal{A}$ and $\mathcal{B}$ of the proposed controller are exhibited in Figure 7.14.

The control performance of the proposed controller $M_{fn}(k)$ and performance indexes of $M_{fn}(k)$ and $U_{open}(k)$ under working conditions $\mathcal{A}$ and $\mathcal{B}$ are given in Table 7.6, in which the data of 1,000 cycles are used to calculated the norm of signals. From Table 7.6, it can be seen that the proposed controller

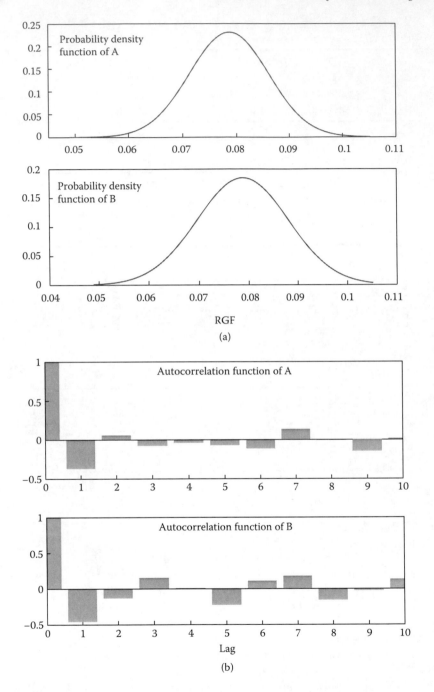

FIGURE 7.12
Performance parameters of RGF. (a) Probability density functions. (b) Auto-correlation functions.

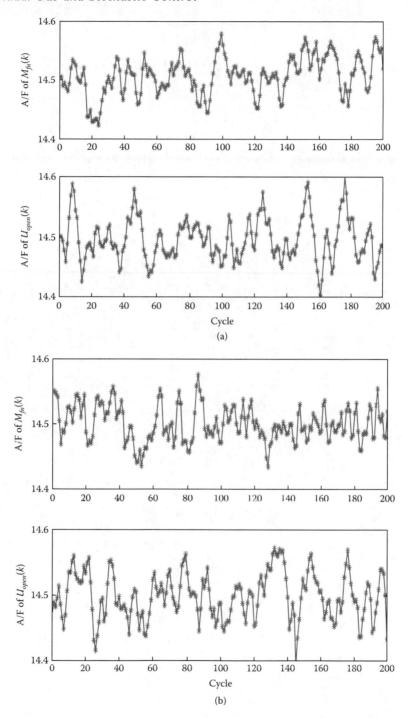

FIGURE 7.13
A/F performance. (a) Case $\mathcal{A}$. (b) Case $\mathcal{B}$.

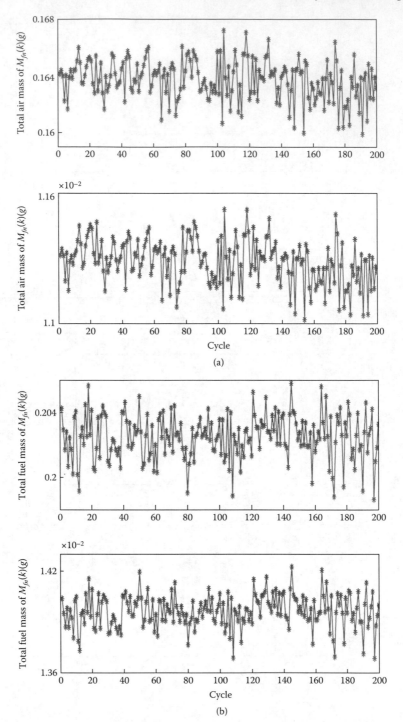

FIGURE 7.14
Total air mass and total fuel mass. (a) Case $\mathcal{A}$. (b) Case $\mathcal{B}$.

TABLE 7.6
Performance indexes of $\mathcal{A}$ and $\mathcal{B}$

Cases	$\sum_{k=1}^{1,000} \|y\| / \sum_{k=1}^{1,000} \|\Delta M_{an}\|$	γ	$J_{M_{fn}(k)}$	$J_{U_{open}}$
$\mathcal{A}$	0.1029	1.4	0.4521	0.5132
$\mathcal{B}$	0.0926	1.4	0.3762	0.4582

$M_{fn}(k)$ achieves robustness in the presence of the cycle-by-cycle fluctuant of the fresh air, and $M_{fn}(k)$ achieves better performance than $U_{open}(k)$, where $J = \sum_{k=1}^{N} (\lambda(k) - \lambda_d)^2$.

7.7 Conclusions

In this chapter, we mainly discussed the statistical property and measurable model of the residual gas for managing the residual gas to achieve the control performance of the engine irrespective of the cyclic variation of RGF.

In fact, in the presented model, the cyclic variation of the RGF was modeled as a Markov process. This stochastic model was motivated by the observation that the combustion state of the current cycle is mainly determined by the combustion state of the previous cycle, and the latter is largely affected by the RGF. A statistical analysis of the experimental samples validated this observation. Based on the obtained dynamic model with this stochastic property, a stochastic optimal control problem and robust L_2-gain disturbance attenuation problem were addressed for the A/F of gasoline engines taking into consideration the cyclic variation of the RGF.

It should be noted that the control schemes require the capability of making in-cylinder pressure measurements. Moreover, the coefficients of the controller need to be tuned for each working condition, since both the state space and the one-step transition probability matrix of the RGF are obtained from the corresponding working conditions. For transient operations, the operation points might be quantified, and the coefficients might be gain scheduled according to the quantified operation conditions.

8

Benchmark Problems for Control and Modeling of Automotive Gasoline Engine

8.1 Introduction

The automotive industry has been confronted by serious social problems, global warming, air pollution, the energy crisis, and traffic safety. Control technologies have been continuously expanded to encounter these issues. That makes powertrain controls highly sophisticated and complex. Therefore, it has been becoming difficult to develop reliability within a reasonable time and with limited resources. In addition, the development period should be shortened to provide new technologies to the market immediately. From the above situation, collaboration between the academic society and the automotive industry has become important. However, the communication between both has been inefficient. It has not been easy for the academic society to know the requirements from the automotive industry. On the other hand, it has been difficult for the automotive industry to know what control technologies are applicable to their needs. Many universities in Japan have no engine test facility, and there has been no good engine model shared among researchers. The situation has prevented university involvement in the area. In order to change the situation, we provided some engine models and benchmark problems of control technologies. In this chapter, three benchmark problems are introduced.

8.2 Benchmark Challenge of Engine Start Control Problem

The SICE Research Committee on Advanced Powertrain Control Theory was established in 2006 to contribute to reinforcing the collaboration between the academic society and the automotive industry. In order to change the situation, we provided a gasoline engine model and a benchmark problem of control technologies. The problem is how to start the engine model as we would an actual engine. Cold engine start is one of the most important problems in automotive engine control because nothing happens before starting the engine. The intermittent phenomena due to the engine cycle appear more strongly at lower engine speed, and almost all of the harmful tailpipe exhaust gases are emitted during the short period after starting the engine. Moreover, stable and smooth engine restart is highly required so that hybrid vehicles (HVs) stop the engine to save fuel at low speeds.

8.2.1   Provided Engine Simulator

Figure 8.1 shows a sketch of the targeted 3.5 L V6 SI engine that has gasoline injection at each intake port. We attempted to make the same situation, which the developers in the automotive industry are confronted with. The provided engine model is constructed to comply with physical principles as much as possible so that the challengers' physical considerations can be effective. Thus, the projection method is applied to the mechanical portion, and the modeling method based on relevant conservation laws is applied to the other portions. The adjustment of the model parameters was determined to minimize the error between the experimental and simulated data. Therefore, the model is a conceptual one.

Figure 8.2 shows the concept of the plant model. Here, we deal with the model for control design as lumped parameter. The universal set of models describes the relationship among the inputs, the disturbances, and the outputs. The physical model is described as one that meets the relevant conservation laws, and the statistical model is described as one that has the parameters adjusted by using the experimental data. The constraint of the conservation laws is relaxed because they are sometimes too strong for

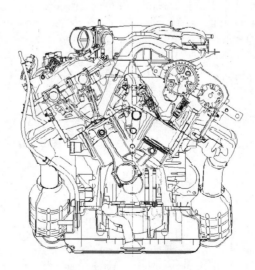

FIGURE 8.1
Targeted V6 SI engine.

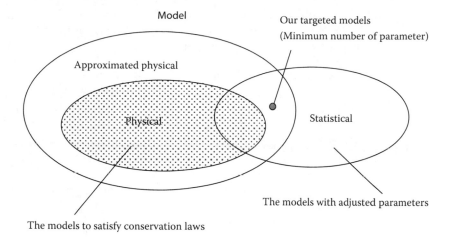

FIGURE 8.2
Concept of plant modeling.

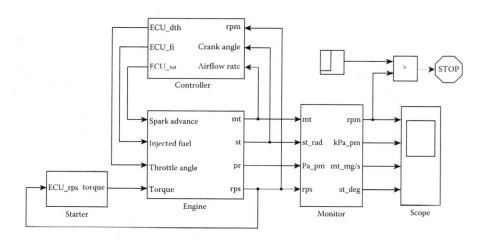

FIGURE 8.3
Top layer of the provided model.

practical usage. It is considered that a relaxed model is similar to the physical model. Thus, it is called approximated physical model. The targeted model for control design is usually in the set of approximated physical models. However, it is very difficult for the challengers to understand the physical background of the approximated physical model. Thus, we decided that the model for the benchmark problem should belong to a set of physical models.

Figure 8.3 shows the top layer of the provided model described with Simulink® [5, 104, 105]. The challengers can access any portions of the model and any model parameters. However, there is no document explaining the detail of each block, although a brief explanation of the model was given at the beginning of the challenges. This situation is quite similar to the one that an engineer given an actual engine encounters. Thus, the challengers must analyze the engine model first. This was our intention. We would like to see their process as well as their control design. Our interests are not only what control they design, but also how they design their control. As shown in Figure 8.3, the provided engine model consists of the engine block, the control block, and the starter motor block. The inputs of the engine are the throttle angles, the amount of the injected fuel mass, and the SA of all the cylinders. The crank angle, the engine speed, and the airflow rate through the throttle valve are fed to the controller. The inputs and outputs of the controller are specified in the control block and cannot be changed by a challenger. The challengers must draw their controllers in the control block. The scope block is added to monitor the engine speed, the intake pressure, the airflow rate through the throttle valve, and the crank angle of cylinder 1.

Figure 8.4 shows the inside of the engine block. It consists of the air dynamics block corresponding to the intake chamber, the cylinder block, and the block for the mechanical portion of the crank system. The throttle valve is directly connected with the atmosphere. The cylinder block describes the right and left banks of the engine, and each bank has three cylinders. The behavior of the fuel in the inlet port and the intake valve of each cylinder is considered. The cylinder pressure profile with the crank angle is calculated, and the airflow rates through the intake and exhaust valves are also calculated with the crank angle. Each exhaust port opens to the atmosphere. The crank shaft is not connected with the transmission. Another version of the model is provided

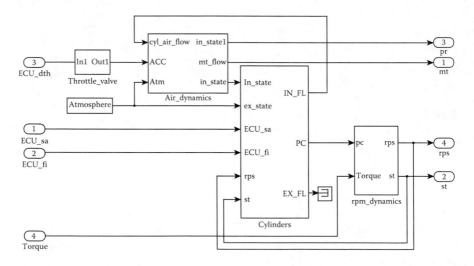

FIGURE 8.4
Inside of the engine block.

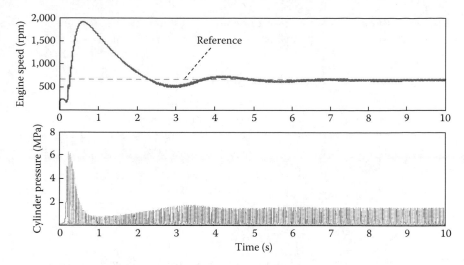

FIGURE 8.5

An example of simulation results: A/F is stoichiometric; SA is constant.

so that the challengers can understand the cold-start engine speed behavior when the throttle angle, the SA, and the injected fuel mass are constant. It can be useful for the challengers to understand the purpose of the designed control in the benchmark problem well.

Figure 8.5 shows an example of simulation results after the engine starts. The upper figure shows the engine speed excursion, and the bottom figure shows the pressure profile of each cylinder. The engine speed fluctuation during each engine cycle appears in the upper figure. The cylinder pressures, the piston inertias, and so on, cause it. Before firing, the starter motor controls the engine at 250 rpm. The air–fuel ratio (A/F), the throttle angle, and the SA are constant. The A/F is stoichiometric. This means the fuel mass in the cylinder is calculated from the induced air mass during the intake stroke as shown by Equation 8.1.

$$\text{Fuel mass} = \frac{\text{Air mass in the cylinder}}{14.5} \qquad (8.1)$$

The engine speed rises rapidly after the ignition, comes down, and gradually converges to 650 rpm. The steep overshoot of the engine speed is the common phenomenon when an engine starts. The other model for validating the robustness of the developed controllers is also provided. The initial crank angle is the top dead center in the previous one. However, it depends on the crank angle when the engine stopped, and it is affected by various factors, for example, the friction torque and the engine speed when the ignition is turned off. The battery voltage changes the engine speed during the cranking period.

The characteristic of the fuel evaporation highly affects the A/F control. The cold-start control must be evaluated in such various conditions.

8.2.2   Benchmark Problem Description

The benchmark problem is to start the engine model and regulate the engine speed at 650 rpm immediately. The requirements are illustrated in Figure 8.6. The dotted line is the reference of the engine speed control. The engine speed must be confined in the area not hatched. The requirements for steady state are the following:

1. The closed-loop is stable.

2. The engine speed converges to 650 rpm.

3. The engine speed reaches 600–700 rpm within 1.5 s.

The requirements for the transient condition are as follows:

4. The overshoot must be sufficiently suppressed.

5. Hunting must not appear.

In the case of Figure 8.6, only requirements 1 and 5 are fulfilled. Figure 8.7 is another example of the evaluations. It fulfills almost all requirements, although the performance for requirement 2 should be improved.

The following requirements for robustness are added. The requirement for steady and transient conditions must be fulfilled with variations of the initial crank angle, friction torque, and fuel evaporation characteristic.

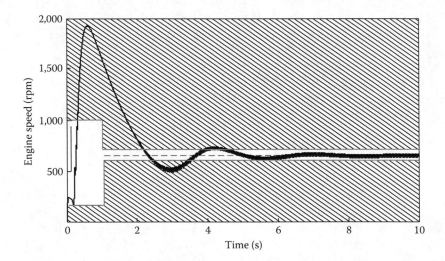

FIGURE 8.6
An example of evaluation.

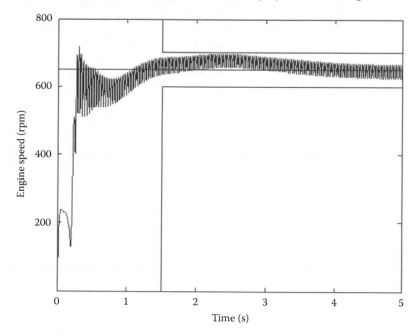

FIGURE 8.7
Another example of evaluation.

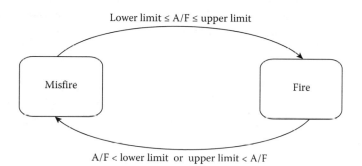

FIGURE 8.8
State transition of engine behavior.

Figure 8.8 shows the behavior that makes starting the engine difficult. The engine model has the state transition between firing and misfiring, as shown in Figure 8.8. The mixture gas in the cylinder ignites when the A/F is in the following range. It does not ignite when the above condition is not satisfied. In the state of firing, the engine is stable, although strong nonlinearity appears. In the state of misfiring, the engine is unstable and reaches the engine stall.

Thus, the key of the engine control is to keep the condition of firing given by Equation 8.2.

$$\text{Lower limit} \leq \text{A/F} \leq \text{Upper limit} \qquad (8.2)$$

Another important feature is the redundancy of the inputs to control the engine speed. The throttle angle, the fuel injection, and the SA affect the engine speed. Thus, the assignment of each role is essential. Usually, the fuel injection is used to control the A/F at the stoichiometric accurate value for exhaust emission, because its torque control range is limited. The throttle angle can control the engine speed in a wide range and is highly economical although the response of the engine speed control is relatively slow compared with the spark advance. The SA is the quickest way but affects the fuel consumption. Thus, the role of each input is as follows in ordinal engine control:

1. Fuel injection: A/F control

2. Throttle angle: Fuel consumption

3. SA: Rapid engine speed control

Figure 8.9 briefly shows a structure of engine control used by the automotive industry when starting an engine. The throttle valve control can be designed as a continuous-time system, but the fuel injection and SA controls must be designed as discrete-time systems. They are actuated every engine stroke. This can be translated as follows.

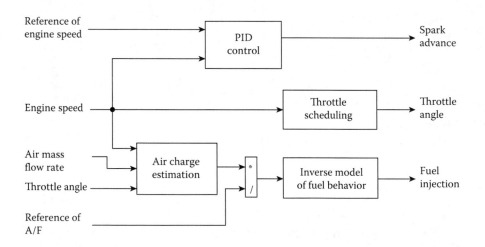

FIGURE 8.9
Ordinary structure of engine control.

The engine model can be described by the following general form:

$$\frac{df(x)}{dt} = f(x, u) \tag{8.3}$$

Here x denotes the state and u denotes the control inputs, and with respect to the crank angle and engine speed, Equation 8.3 can be rewritten as

$$\frac{dx}{d\theta} = \frac{f(x, u)}{\omega} \tag{8.4}$$

Using Equation 8.4, the engine model can be transformed to the discrete crank angle system. However, this may be an issue at lower engine speeds. This means that we do not have frequent occasions to actuate the engine. This is one of the reasons for the difficulty in suppressing the engine overshoot. Thus, we have to rely on the feedforward for the purpose.

Figure 8.10 shows a possible control strategy. It changes with time. This is another important feature of this problem, although someone can take a time-invariant control strategy. But, the time variant control strategy has bases from physical phenomena. The discrete-event feature exists behind the scene. Thus, it can be considered that the strategy is natural and fundamental for the problem.

From the above discussion, the distinctive features of this benchmark problem are summarized below:

1. State transition between stable and unstable states

2. Redundancy of inputs

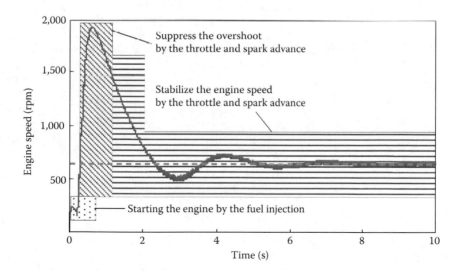

FIGURE 8.10
The strategy of engine control.

3. Continuous-time and discrete-event system

4. Time-variant control strategy

5. Combination of feedforward and feedback control

Thus, it can be considered that this problem is complex. Here, complex means that it cannot be solved by only one method and a combination of methods is required. For example, a few advanced controls were applied to idle speed control [106]. The control can be derived from the framework of the linear control design. Therefore, idle speed control is simple in this context. However, this problem cannot be solved by only one framework and requires combining other methods. An engineer of the automotive industry is struggling to deal with this type of complexity. On the other hand, simple things are piled in the education of universities. The direction is from simple to complex. But, as mentioned above, an engineer of the automotive industry is given a complex problem at the beginning. He must take the inverse direction, from complex to simple. This is the reason why we adopted this problem.

Figure 8.11 shows an example of control results with the control structure shown in Figure 8.9 according to the strategy shown in Figure 8.10. This result meets the requirements of the problem, although a little improvement is still possible. However, the problem is still difficult because robustness to the variations of fuel, engine setting, and environment is highly required. To guarantee reliability in the marketplace is a tough problem. Usually, it takes a lot of time to complete it. Moreover, the requirements from the reliability and the

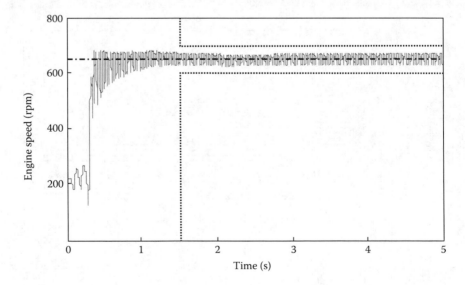

FIGURE 8.11
An example of a control result.

exhaust gas emission control cause the trade-off issue. Thus, more efficient and accrete control designs are highly required.

8.2.3   Approaches of Challengers

Seven challengers reported their intermediate results in the 36th SICE Symposium on Control Theory in Sapporo, Japan, on September 6, 2007 [111–114]. All challengers succeeded in starting the engine, although some levels of the overshoot and engine speed stability were seen in their results. Their approaches can be classified into the following classes.

A: Analyzing the model to know the physical background and recreating the equations of the model

B: Using a control design similar to the traditional approach with an advanced control methodology applied to a portion

C: Trying to find the optimal input time sequence profiles and the combination with the feedback control around the profiles

D: Using a simulation data-based approach combined with the traditional engine control structure

E: Using a reckless approach to use the state of misfire

We highly appreciate the efforts of the challenger A types, who analyze the engine model. It must be very hard work. They know that the engine model is very complex and has a discrete-event feature and strong nonlinearity. They could not find a theory directly applicable to the set of equations. They try to simplify the equations or find a good way to deal with the model from computer science and information theory.

The challenger B types adopt a control architecture that is similar to the existing one. That would be the most promising. The physical consideration seems to be very useful. Their advanced control seems to be able to improve the performance compared with the existing one and also seems to demonstrate the power. One of them analyzed the measurement and actuation timings very carefully. That is very important in control design.

The challenger C types are still struggling to get a good result, although they succeeded in starting the engine and stabilizing the engine speed. The optimization may be caught by a local minimum and not reach the global optimal input profile. They have a problem when the feedback starts because it does not work well during the condition of misfiring. The approaches are very attractive because they can clarify the optimal operation of the throttle, the fuel injection, and the SA and can be expanded to the case in which more constraints are added, for example, warming up the catalyst and minimizing the HC exhaust gas emission. They tried to manually optimize the fuel injection and SA profiles under the condition of the constant throttle angle. Other challengers obtained the three input profiles at the same time by a numerical optimization. The challengers are aware of the need to start their feedback from a suitable timing, and they intend to start the engine with only the feedforward. They would add some feedback methods to the feedforward. But nobody can succeed at this moment.

The challenger D types applied a data-based approach to the air charge estimation. The result looks good. But, it is difficult to understand the physical background because their method is quite different from the ordinal one. We need the analysis why it is good and will watch the progress. This method could provide the way to systematically enhance the knowledge to manage the control design. The engineers in the automotive industry would be heuristically working on the same purpose with many experiments.

The idea of the challenger E types is out of our intention. The other challengers try to keep the situation of firing, but the challenger E types actively use the condition of misfiring. They give a stimulus to the industry. Their idea is connected with the variable cylinder number operation considered as a way to increase the fuel economy. Definitely, it is a way to suppress the engine speed overshoot. However, it has many issues not described here. However, they can be resolved by continuous efforts in the future.

All agreed that the engine model is complex. They observed hybrid features and strong nonlinearity in the model equations. One said that no existing model-based control theory can deal with such a complex system. It may be too early to let this be the final conclusion because control theories have contributed to the current prosperity of the automotive industry, although almost all of the actual applications have such complex features. Thus, this type of discussion is very important.

Many of the challengers try to get useful information from the engine model. One way is to clarify the model equations, and the other is to make a simple model from the simulation data numerically or the physical consideration. This is highly connected with model simplification, which means the reduction of the system order and number of parameters. It is obvious the structure identification is highly required for the purpose. It can also be obvious that the connection with physical considerations is effective.

A key to succeed in control design seems to be the control architecture, as shown in Figure 8.12. The portion of the fuel injection control plays the role of keeping the A/F ignitable. It is difficult to control the engine speed when the state transition between firing and misfiring happens. The challengers who succeeded in starting the engine commonly used the architecture. Thus, it is obvious that the architecture makes the control problem easier. Once the fuel injection control is successfully developed, the control design of other portions can become relatively easy. Almost all are trying to know the optimum input time sequence profiles. Some took trial-and-error approaches, and the others took numerical optimization methods. Even for the trial, the architecture is useful.

Quite a few of the challengers are looking into MPC. One already considered an observer for the purpose and did a preliminary study of MPC. It may play an important role in this research activity.

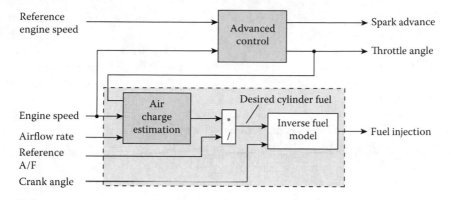

FIGURE 8.12
Common architecture of a successful control.

Information theory and methods from computer science may be useful to attack the complex system, although we did not discuss what types of new methods are possible.

A carefully worked-out plan is also essential. It is a natural way to encounter a complex problem. The requirements analysis may be the first step of the control design, and the analysis of the model should be the second step. However, there are challengers who try a specific control theory that may not cover all of the requirements before getting the results of the above steps. They will fail because we designed the benchmark problem to not be solved in such a way. They should make it clear first why it can fulfill the requirements. Actually, PI control of the throttle angle can regulate the engine speed at 650 rpm, although the response is not quick and it cannot suppress the overshoot sufficiently. However, the advanced control without the required careful consideration may be worse than a PI control combined with other methods obviously fulfilling requirements. We think the combination of various methodologies is required for the benchmark problem, and the optimization of the combination is not easy. The engineers in the automotive industry struggle to resolve such problems.

From the above, we have to consider the purpose of advanced control. It may provide more accurate control, more robust control, and more systematic control design than the existing control designs. It should also be expected to cover more complex control design than before.

8.2.4   Summary

The SICE Research Committee on Advanced Powertrain Control Theory provided a V6 gasoline engine model and the control problem starting the engine. Seven challengers reported their intermediate results on September 6, 2007, in the 36th SICE Symposium on Control Theory. There were more groups working on the engine model, although we did not get their results. Actually, there are groups that want to join this activity in the world. We feel that this activity is very useful for academic parties and the industry. The problem is complex, which means not solved by only an existing control methodology, and requires a combination of methodologies. We are watching not only the control results, but also the control design process. Academic parties take various approaches, and almost all already succeeded in starting the engine but the overshoot of the engine speed still appears and robustness is still a problem. We can expect further progress and hope to apply their controls to an actual engine.

8.3 Benchmark Problem for Nonlinear Identification of Automotive Engine

Toyota, Nissan, Honda, Mazda, DENSO, and Hitachi established the Japan Calibration User Group (JCUG) to initiate the efficient engine calibration environment in April 2009. Engine calibration means the final coordination of the developed electronic control unit (ECU) mainly by adjusting and properly determining parameters of developed engine control. A typical gasoline engine with an intake variable valve timing system has about 15,000 parameters decided by experiments. Due to the strong demands for CO_2 reduction, environment protection, and vehicle safety, the complexity of engine control has been progressing. It has become such a big issue that the required experiments are rapidly and exponentially increasing. To encounter the issue, the automotive industry has introduced model-based calibration (MBC) into their calibration processes [111]. MBC has been successful in the steady-state calibration area. According to the success, the automotive industry is extending the usage to the transient calibration area. For both steady-state MBC and transient MBC, it is important to construct engine models that satisfy the required accuracy by reasonable effort. From this point of view, the Joint Research Committee on Automotive Control and Modeling of JSAE and SICE released "Benchmark Problem for Nonlinear Identification of Automotive Engine" in 2011.

8.3.1 Provided Engine Model

This chapter describes important equations of the engine model and the implementation on Simulink. That can be useful for challengers to understand automotive engines.

Deriving Model Equations. The provided engine model is not a mean-value model but an in-cycle model for an automotive L4 gasoline engine. Figure 8.13 briefly shows an intake system of an automotive engine.

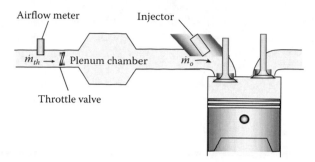

FIGURE 8.13
Brief sketch of intake system.

The modeled engine has four cylinders, although Figure 8.13 simply describes only a single cylinder. The gas composition is a mixture of HC, CO_2, CO, O_2, H_2, H_2O, and N_2. The model is derived from the molecules and the energy conservation laws with no modification by using experimental data so that challengers can guess the engine model behavior with their physical considerations. Therefore, the model well describes the engine behavior qualitatively but not quantitatively.

The air charge estimation is the most fundamental function of current engine control since other controls, such as the SA, fuel injections, and other actuator controls, are determined according to the result. To estimate the air charge, the following equations are generally used:

$$\frac{d\rho_m V_m}{dt} = \dot{m}_{th} - \dot{m}_o(\omega, p_m, \cdots) \tag{8.5}$$

$$\frac{d}{dt}\left(\frac{1}{\kappa - 1}p_m V_m\right) = \frac{\kappa}{\kappa - 1}\dot{m}_{th}RT_a - \frac{\kappa}{\kappa - 1}\dot{m}_o(\omega, p_m, \cdots)RT_m \tag{8.6}$$

where ρ_m denotes the intake air density. To eliminate ρ_m from the above equation, the ideal gas state equation

$$p_m = \rho_m RT_m \tag{8.7}$$

is applied. Therefore, the second-order system with p_m and T_m is obtained. The flow rates m_t and m_c derive the molecules and energy flows depending on the upper and downstream conditions. That allows us to calculate the gas composition and the internal energy in the plenum chamber. (Equation 8.6 is a simplified expression. The conservation law of numbers of molecules is considered in the actual calculation.) The same equation to calculate the airflow rate is applied to the intake and exhaust valves, as well as to the throttle valve. The energy conservation law applied to the cylinders becomes

$$\frac{d}{dt}\left(\frac{1}{\kappa - 1}p_{c,i}V_{c,i}\right) = \frac{\kappa_{in,i}}{\kappa_{in,i} - 1}m_{in,i}RT_a - \frac{\kappa_{ex,i}}{\kappa_{ex,i} - 1}m_{ex,i}RT_{c,i} - p_{c,i}\frac{dV_{c,i}}{dt} \tag{8.8}$$

where $m_{in,i}$ is the gas flow rate through the intake valve and $m_{ex,i}$ is the gas flow rate through the exhaust valve. The third term of the right-hand side indicates the energy flow from the cylinder gas to the piston. $\kappa_{in,i}$ and $\kappa_{ex,i}$ are the specific heat ratios of the gases through the intake and exhaust valves that depend on the gas compositions and the directions of the flows.

The accumulated heat energy generated by the combustion is expressed by

$$\frac{Q_{i,k}}{Hf_{i,k}} = 1 - \exp\left\{A\frac{(\theta - \theta_{s,k})^{M+1}}{\Delta\theta}\right\} \tag{8.9}$$

where $Q_{i,k}$ is the accumulated heat energy of cylinder i at engine cycle k, $f_{i,k}$ is the amount of the induced fuel into the cylinder, and θ_s is the SA. H, A, M, and $\Delta\theta$ are constant. That means that the profiles of combustion do not change regardless of engine operation conditions. The cylinder gas is divided into burned and unburned gas portions during the combustion. It is supposed that the pressures are equal to each other.

It is considered that the unburned gas is transformed to the burned gas defined by Figure 8.14. The generated heat is calculated according to the transformed gas composition. The composition shown in Figure 8.14 is calculated by the execution of the M-file, engine_sim_set.m, setting the required parameters. In the plenum chamber, the exhaust manifold, and the cylinders,

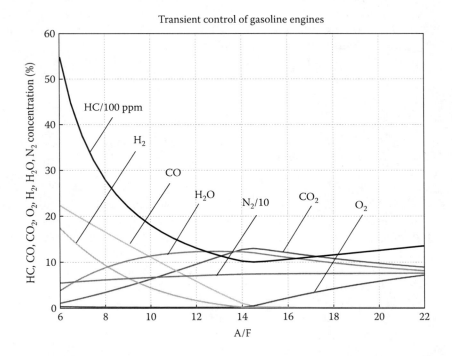

FIGURE 8.14
Gas composition of burned gas.

the molecules of HC, CO_2, CO, O_2, H_2, H_2O, and N_2 are mixed. Thus, the concentration of each component in the plenum chambers and the cylinders is dynamically changed.

The instantaneous indicated torque for cylinder i is calculated by

$$\tau_{e,i} = p_{c,i} \frac{V_{c,i}}{d\theta} \tag{8.10}$$

but it fluctuates with the crank angle according to the change of the cylinder volume and the cylinder pressure. Thus, the mean indicated torque during the engine cycle is adopted.

Implementation of Engine Model. The provided engine simulator works on Simulink and Stateflow. The challengers can access all blocks and M-files used in the model. Thus, the challengers can analyze the model in detail with no restrictions. However, the challengers must not change the inside of the engine_4cyl block, although they can revise the insides of the controller and the calibration system blocks freely, except for the inputs and outputs.

Figure 8.15 shows the top layer of the engine model. It consists of four major blocks, such as the engine_4cyl, the controller, the calibration system, and the monitor blocks. When you click on the engine_4cyl block, you can look into the inside of the blocks. Figure 8.16 shows the inside of the engine_4cyl

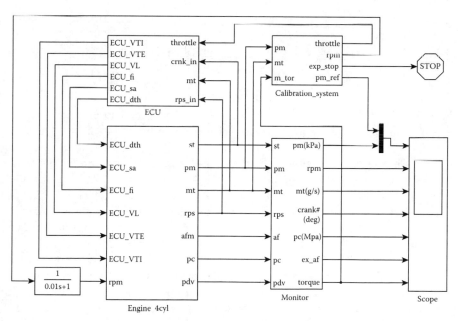

The provided engine model for the Benchmark Problem 1
Joint Research Committee on Automotive Control and Modeling of JSAE and SICE

FIGURE 8.15
The top layer of the provided engine simulator.

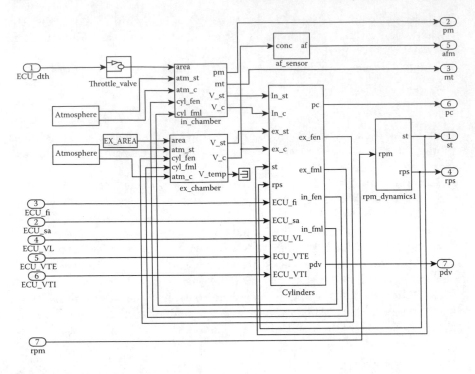

FIGURE 8.16
The second layer: the engine system consisting of the intake system, the
cylinders, and the exhaust system.

block. The cylinder, input chamber (corresponds to the plenum chamber), and
exhaust chamber (corresponds to the exhaust manifold) blocks are shown in
this figure.

Figure 8.17 shows the inside of the cylinder block. There are four cylinder
blocks, from cylinder 1 to cylinder 4, in this figure. The challengers can look
into the detail of each cylinder block when they click on it. Like this, the
challengers can access all blocks but are not allowed to change any blocks.

The calibration system block manages how to acquire the simulation data
and change the engine simulation condition. Figure 8.18 shows simulation
data acquisition flow in the calibration system block that corresponds to the
measurement data acquisition in the real world. The block manages the simulation data acquisition in which the engine speed and the intake pressure
are changed as the defined schedule. It is supposed that the engine speed is
accurately controlled with the engine dynamometer, but the intake pressure
is controlled with ECU. Thus, the calibration system feeds the throttle angle
signal to the ECU according to the error between the intake pressure and the
reference. The measurements (simulation data acquisitions) must be carried
out at stable engine conditions. Therefore, the role of the calibration system is

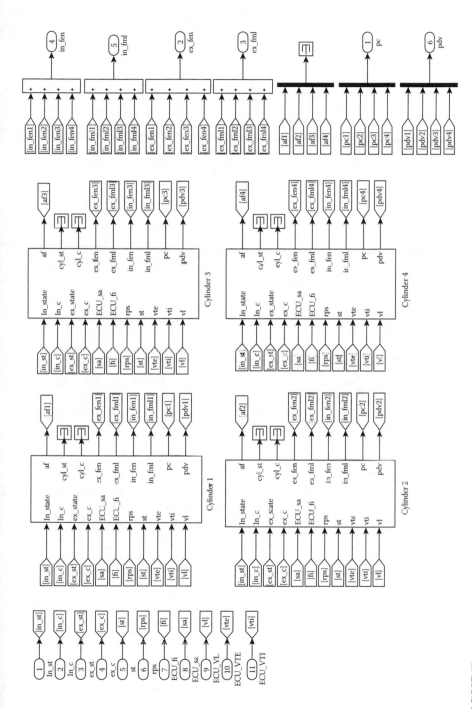

FIGURE 8.17

The third layer: from cylinders 1 to 4.

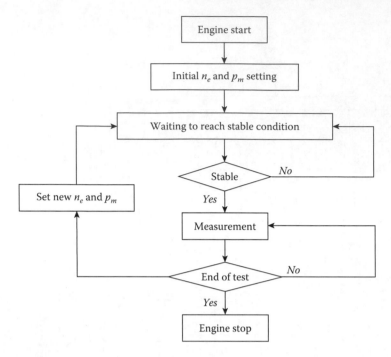

FIGURE 8.18

Data acquisition flow in the calibration system block.

to control the intake pressure at the reference value, check if the error between the intake pressure and the reference becomes sufficiently small, and measure the airflow rate and mean indicated torque.

Figure 8.19 shows a simulation result by using the data acquisition flow shown in Figure 8.18. The top figure shows the intake pressure and the reference, the second figure is the engine speed, the third figure is the airflow rate through the throttle valve, the fourth figure is the crank angle, the fifth figure is the cylinder pressures, the sixth figure is the A/F, the seventh figure is the flag indicating the measurement is ready, and the bottom figure shows the control error of the intake pressure. The intake pressure was controlled by PI control in the calibration system block. A big error is seen after changing the reference, as shown in the bottom figure. The intake pressure converges to the reference value gradually. Thus, the measurement must be carried out after the intake pressure control error becomes sufficiently small. The value 1 of the seventh figure means that the control error is sufficiently small, and 0 means that the measurement is not ready. It takes a little bit of time for the A/F to reach the stable condition, as shown in the sixth figure. Thus, the measurement should wait a little bit after the flag turns to 1. After a measurement, the calibration system block changes the engine condition to the new engine speed and the intake pressure.

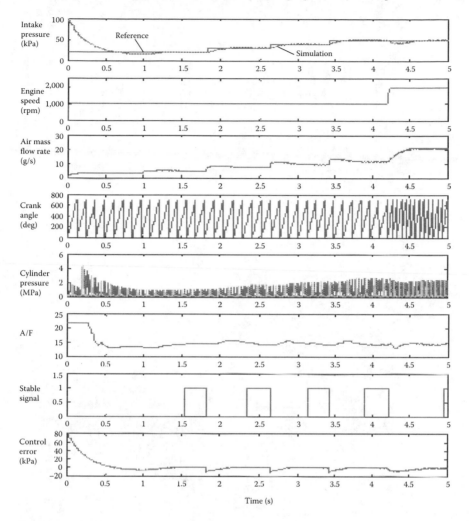

FIGURE 8.19
Measurement process simulation.

Figure 8.20 shows the detail of the intake pressure profile. The oscillation of the intake pressure is caused by the intake stroke. A similar fluctuation can be found in the engine speed in an engine cycle. The benchmark problem does not require identifying the oscillation. This indicates a filtering method to eliminate the necessity of the oscillation.

The engine speed, the intake pressure, and the airflow rate were accumulated to the workspace. That corresponds to the automated measurement system. Figure 8.21 shows the accumulated simulation data of the airflow rate through the throttle valve. The measurements are usually done in all possible engine operation conditions.

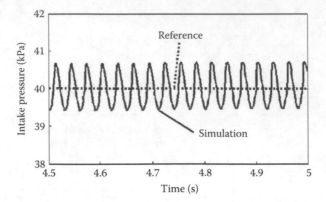

FIGURE 8.20
Oscillation of the intake pressure.

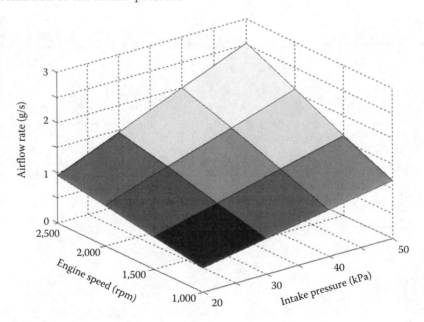

FIGURE 8.21
Accumulated simulation data of airflow rate.

8.3.2 Benchmark Problem

The provided benchmark problem is described as follows:

1. Engine operating condition

   - Engine speed: 500–3,000 rpm
   - Intake pressure 20–95 kpa

- Intake valve phase: 0–60° crank angle
- Exhaust valve phase: 0–60° crank angle
- Intake valve lift: 2–6 mm

2. Actuated signals

    Engine speed, throttle opening angle, variable intake and exhaust valve phases, and variable intake valve lift

3. Identified signals

    - Airflow rate through the throttle valve
    - Mean indicated torque

      Both data are accumulated in the calibration system blocks. In the identification, they can use any data in the controller block and the calibration system block, but must not use data from the other portions.

4. Evaluation condition

    - Static condition: At the following grids.
      Engine speed: 500, 1,000, 1,500, 2,000, 2,500, and 3,000 rpm.
      Throttle angle: 20, 35, 50, 65, 80, and 95 kPa.
      Intake valve phase in crank angle: 0°, 20°, 40°, and 60°.
      Exhaust valve phase in crank angle: 0°, 20°, 40°, and 60°.
      Intake valve lift: 2, 4, and 6 mm.

    - Transient condition: 500 points are sequentially selected from the above static conditions randomly, and the engine condition moves from a point to another point with randomly determined time steps (minimum, 100 ms; maximum, 2 s).

    - Evaluation of accuracy: For static and dynamic evaluations, the mean average of the square root error and the absolute error between simulation data and the identified model are evaluated. The identified model must be one. That means that human interaction is not allowed to switch the static and dynamic models. The challengers must notice that the simulation error, not the one-step prediction error, is evaluated. That means challengers must use simulation data in every prediction.

    - Evaluation of measurement effort: The static and dynamic measurement efforts are evaluated.
      Static condition: The number of simulation points.
      Transient condition: The data length.
      This means that the challengers must determine simulation conditions so that the above data are minimum.

8.3.3   Sample Results

By using the data, an empirical model was created by, for example, the least-squares method. A polynomial function, radial basis functions, and a piecewise affine model are often used. In this chapter, the following second-order polynomial model was applied:

$$\dot{m}_{th}(k) = \alpha_1 \omega^2(k) + \alpha_2 \omega(k)(k) p_m(k) + \alpha_3 p_m^2(k) + \alpha_4 \qquad (8.11)$$

Figure 8.22 is the comparison of the accumulated data and the empirical model expressed by (8.11). The horizontal axis indicates the measurement time. The correlation is quite good because the simulation was done only in the small area of the engine operation condition, and many of important phenomena were ignored in the provided engine simulator. This benchmark problem may not be easy because the intake and exhaust valve timing system and the intake valve lift system must also be considered.

To identify transient models, we may use data acquisition work flows similar to the one shown in Figure 8.18. A challenger may add small perturbation of the throttle angle after reaching the stable condition and apply the linear identification to the measured data. In this way, he can get a set of the local affine models and integrate them to obtain the global nonlinear model.

Figure 8.23 is an example of a data set to create a transient model. The throttle angle sweeps the admissible opening angle with a small perturbation. The bottom figure shows the response of the intake pressure. Off course, it is difficult to do so in actual experiments. For example, rapidly changing the engine condition causes a big fluctuation of A/F, and that can cause misfiring. It highly affects engine behaviors. In this benchmark problem, rough excitations can be allowed. But, challengers should notice that rough excitations might excite hidden modes of the simulator and make the identification difficult.

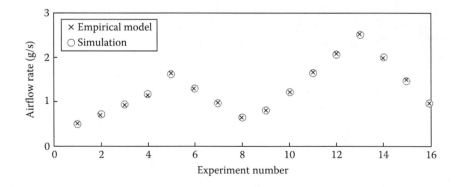

FIGURE 8.22
Comparison of accumulated data and empirical model.

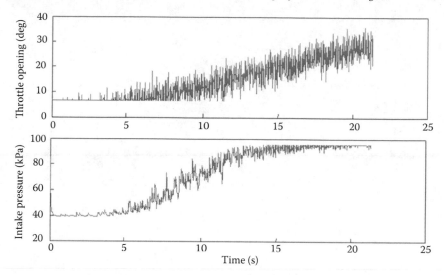

FIGURE 8.23
A simulation data set for identification.

Local affine models can be identified from the data at each intake pressure. The global nonlinear model can be estimated from the local model. Equation 8.12 is an example of the nonlinear model from the throttle angle to the intake pressure. A difficulty of the identification appears when the intake pressure approaches the atmospheric pressure because the numerical calculation of a model tends to be unstable around that pressure.

$$p_m(k+1) = \left\{ \alpha_1 p_m^{-0.24}(k) + \alpha_2 p_m^{-0.12}(k) + \alpha_3 \right\} p_m(k)$$
$$+ \left\{ \beta_1 p_m^{-0.002}(k) + \beta_2 p_m^{-0.001}(k) + \beta_3 \right\} \theta(k) + \gamma \tag{8.12}$$

Figure 8.24 shows the comparison of the identified nonlinear model and experimental data. From Figure 8.24, it is obvious that the model is nonlinear because the response changes with the amplitude of the intake pressure.

8.3.4   Challengers' Results

Seventeen people downloaded this benchmark problem. Two selected papers were reported in IFAC AAC (Advances in Automotive Control) 2013 in Tokyo.

8.3.5   Summary

The Joint Research Committee on Automotive Control and Modeling of JSAE and SICE released "Benchmark Problem for Nonlinear Identification of Automotive Engine" in 2011. The challengers are requested to generate data for identification from the provided engine simulator and develop a simple model

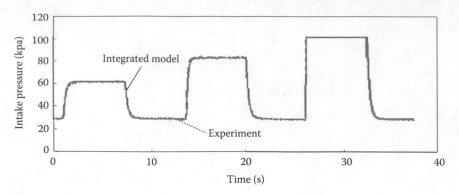

FIGURE 8.24
Transient model of the intake pressure.

of the indicated torque and airflow rate through the throttle valve that can accurately re-create simulation data on all operation conditions. The accuracy and data size used in the identification are evaluated. That means the efficient design of simulations is highly required. The challengers' results were shared in IFAC AAC (Advances in Automotive Control) 2013 in Tokyo.

8.4   Benchmark Problem for Boundary Modeling and Near-Boundary Control

One of the current driving forces of engine development is CO_2 reduction. It has pushed engine operation to the area where misfiring and knocking easily happen. Combustion with a lean A/F and high exhaust gas recirculation (EGR) ratio can reduce pumping loss, but such directions increase the risk of misfiring. A high compression ratio increases the combustion efficiency, but it also increases the risk of knocking. A high boost pressure also increases the risk of knocking. Misfiring may damage the catalyst seriously, and knocking makes passengers uncomfortable and may cause significant engine damage. Generally, optimal fuel efficiency design tends to cause such issues because the optimality is sometimes obtained on the boundary of the admissible domain. In other words, highly optimal control is pushed toward the boundary between normal and abnormal engine operations. That is the reason why boundary modeling and sophisticated near-boundary controls are necessary for modern engines. Considering this situation, the Joint Research Committee on Automotive Control and Modeling of JSAE and SICE released "Benchmark Problem for Boundary Modeling and Near Boundary Control" in 2014.

8.4.1 Provided Engine Model

Figure 8.25 shows the high level model description (HLMD) [112] of the provided engine model. The model consists of the intake system, the cylinder gas, the exhaust system, the mechanical rotational system, the cylinder wall, and the atmosphere. The fresh air flows into the intake system from the atmosphere through the throttle valve and is mixed with the backflow gas from the cylinder to the intake system. The mixture gas is induced into the cylinder while the intake valve is open. The exhaust gas flows out to the exhaust system and the atmosphere. It is supposed that the composition, the pressure, and the density of the atmosphere are constant because the atmosphere is sufficiently massive. The cylinder gas is divided into burned and unburned gas portions because the combustion is modeled so that the flame is propagating from the burned gas to the unburned gas portions. During the combustion, we have assumptions that the pressures of both portions are equal to each other, and the summation of the volumes is equal to the chamber volume. They are shown in the block in Figure 8.25, which we call the constraint block, connected with the cylinder gas by the dotted lines, and each number of the atoms is conserved and the burned gas composition, pressure, temperature, mean gas constant, and degree of freedom change according to the progress of combustion. The generated heat is accumulated in E_b and transferred to E_u because both

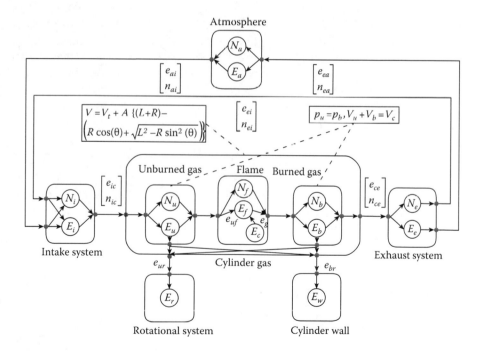

FIGURE 8.25
HLMD of engine model.

volumes change during the combustion. That effect is represented by e_{ub}. The energy accumulators of E_b and E_u also exchange energies with E_m and E_w, which are the mechanical energy and the heat of the cylinder wall, respectively. The cylinder volume is represented in the other constraint block as shown in Figure 8.25.

In Figure 8.25, E_* means the amount of energy of the component, and N_* indicates the molecules of HC, CO_2, CO, O_2, H_2, H_2O, and N_2. Therefore, E_* is a scalar and N_* is a vector. The suffix $*$ indicates the components of the engine. The solid lines indicate the energy and molecule flow vectors, and the directions of the arrows show the sign convention. The attached small letters with the lines indicate the names of energy flow and molecule flow vectors represented by e_{**} and n_{**}, respectively. The first letter of the suffix indicates the source, and the following letter indicates the destination.

The model is derived by the conservation laws represented by

$$\frac{d}{dt}\begin{bmatrix} E_i \\ N_i \end{bmatrix} = \sum \begin{bmatrix} e_{*i} \\ n_{*i} \end{bmatrix} - \sum \begin{bmatrix} e_{i*} \\ n_{i*} \end{bmatrix} + \begin{bmatrix} e_{ig} \\ n_{ig} \end{bmatrix} \tag{8.13}$$

where e_{ig} is the generated energy flow, and n_{ig} indicates the generated molecule flow vector. The molecules are conserved in the components in which the chemical reactions happen. In Figure 8.25, the energy and molecules are not accumulated in the flame block, but molecule flows between the compositions of the inlet and outlet flows are changed according to the conservation laws of the atom. The energy flow e_{uf} transferred to the burned gas by the pressure must be considered. It is supposed that E_f and N_f are always zero.

The composition of burned gas according to the A/F is calculated as shown in Figure 8.14. The combustion speed is represented by the Wiebe function [104], of which the parameters are changed with the engine speed, air charge, fraction of EGR, and phase shift of the intake variable timing system. The Wiebe function is represented by

$$Q = Q_{max}\left\{1 - \exp\left(A\frac{\theta - \theta_{sa}}{\Delta\theta}\right)^{m+1}\right\} \tag{8.14}$$

In Equation 8.14, Q denotes the accumulated heat, Q_{max} denotes the amount of generated energy due to the combustion, and A, $\Delta\theta$, and m are parameters depending on the ω, η, λ, μ, and θ_{sa}, where μ denotes the phase shift angle of the variable intake valve timing.

θ_{50} is defined as θ in (8.14), giving $Q = 0.5$. It means the crank angle at which half of the mixture gas has been burned. The change of θ_{50} with the engine operation condition is represented by

$$\theta_{50} - \theta_{sa} = \exp(3.267 + 0.001722\omega - 0.6367\eta + 0.5782\lambda + 0.1459\mu - 0.8377\theta_{sa}$$
$$+ 1.680\theta_{sa}^2 - 0.002677\omega\theta_{sa}) \tag{8.15}$$

The engine simulator is constructed such that the conservation laws of the energy and the molecules are satisfied as much as possible, but the empirical model of combustion is combined with the physical engine model portions. The reason why portions are almost physical models is that the challengers can estimate the behavior with physical consideration. Therefore, the engine model is relatively simple but highly nonlinear, and the order is more than 50. It is not expected that the model recreates experimental data accurately, although it describes engine behaviors qualitatively. Especially, the ignition failure is not modeled, and the definition of misfiring in this benchmark problem means the event where the exhaust valve opens before the combustion is completed.

The amount of fuel injection is determined by the estimation of the O_2 flow rate through the intake valve so that the intake gas flow satisfies (8.16).

$$n_{HC}CH_y + \frac{2n_{HC} + n_{H_2O}}{2}O_2 \rightarrow n_{HC}CO_2 + n_{H2O}H_2O \qquad (8.16)$$

The amount of fuel injection is determined at the end of the intake, but the fuel injection is completed before the intake stroke. Thus, the A/F is not controlled at the targeted value 14.5 on transient conditions, although it is well controlled on static conditions that mean all actuation inputs are constant.

Table 8.1 shows the differences among the models provided for the benchmark problems. Big differences are that the combustion profile changes with the engine operation condition and the knock and misfiring models are implemented in the model in this chapter.

8.4.2   Benchmark Problem

To achieve the near-boundary control, we have to take two steps. In the first step, the boundaries of misfiring and knocking are identified. In the next step, a near-boundary control is designed. Thus, this benchmark problem is divided into two portions.

Benchmark Problem 1: Identification of Boundary Models. The model includes two features relating to the boundary model. One is the boundary of misfiring and the other is the one of knocking. In this benchmark problem, misfiring is defined as the event that the exhaust valve opens before the combustion is completed in each cylinder, as shown in Figure 8.26. The mixture gas unburned by when the exhaust valve opens flows out to the exhaust system, and the fuel is observed as the HC exhaust emission. Such fuel does not contribute the indicated torque. Thus, the effective torque decreases due to the effect.

Knocking is defined as the event that the knocking integral represented by

$$\int_{t_{is}}^{t} \frac{1}{\varphi_1 p_c(t)^{\varphi_2} \exp\left(\dfrac{-\varphi_3}{T_c(t)}\right)} dt \qquad (8.17)$$

reaches 1 before the end of combustion [113], where T_c is the unburned gas temperature and φ_1, φ_2, and φ_3 are constant parameters.

Knock integral is calculated in the engine simulator as shown in Figure 8.27, and the remaining mixture gas of the combustion burns in an

TABLE 8.1

Differences from the relevant previous models

	Model in Section 8.2 (Figure 8.3)	Model in Section 8.3 (Figure 8.15)	Model in Section 8.4 (Figure 8.25)
Numbers of cylinder	6	4	4
Gas composition	Fresh gas + burned gas	HC, CO_2, CO, O_2, H_2, H_2O, N_2	HC, CO_2, CO, O_2, H_2, H_2O, N_2
Fuel injection	The amount fed by the controller	A/F 14.5 is targeted	A/F 14.5 is targeted
Engine speed	Calculated	Given	Given
Engine outlet gas	To the atmosphere	To exhaust manifold and pipe	To exhaust manifold and pipe
A/F sensor	Not implemented	Implemented at the exhaust manifold	Implemented at the exhaust manifold
Combustion profile	Fixed	Fixed	**Modeled**
EGR and Misfiring	Not implemented	Not implemented	**Implemented**
Knocking	Not implemented	Not implemented	**Implemented**

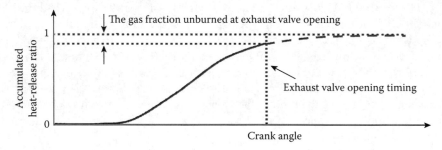

FIGURE 8.26

Definition of misfiring.

instant when the knocking integral reaches 1. Therefore, it corresponds to the knocking intensity.

We recommend challengers tackle the misfiring boundary identification first because the boundary is almost determined by the combustion profiles, which are calculated from the engine speed, air charge, phase shift of VVT,

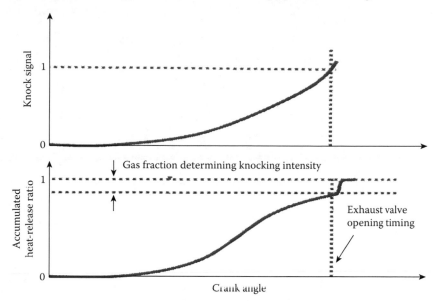

FIGURE 8.27
Definition of knocking.

fraction of EGR, and A/F. In the simulation, A/F is not constant, and it is known by the execution result of the engine model. However, the consideration with neglecting the dynamics of A/F gives a good perspective of the boundary. On the other hand, it is not easy to find such a static function strongly connecting with the knocking boundary effected by the A/F dynamics, and so on.

Figure 8.28 shows what the knocking boundary looks like. Early setting of SA and higher setting of intake pressure make knocking tend to happen as shown in Figure 8.28. Thus, the operation region with the SA and the intake pressure is divided into regions in which knocking happens and does not. The engine operation condition is represented by

$$x = [p_m, \theta_{sa}]^T \qquad (8.18)$$

The boundary is represented by

$$h(x) = \begin{cases} < 0, & \text{knocking happens} \\ = 0, & \text{the boundary} \\ > 0, & \text{knocking does not happen} \end{cases} \qquad (8.19)$$

Therefore, the identification of the boundary model means determining the function h. In the benchmark problem,

$$x = [p_m, \theta_{sa}, egr, evt]^T \qquad (8.20)$$

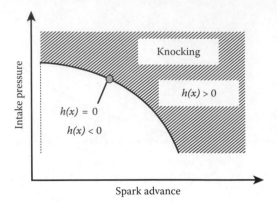

FIGURE 8.28
Example of knocking boundary.

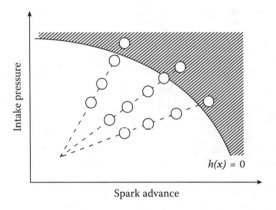

FIGURE 8.29
Example of boundary modeling strategy.

where *egr* is the fraction of EGR and *evt* the cam phase shift of the intake variable valve timing system.

Figure 8.29 shows an example of the boundary modeling strategy. The circles in the figure indicate experiments to know whether knocking happens or does not. It is not difficult to identify h, but the engine operation region is the 5D space in this case. Thus, the task to obtain a sufficient simulation data set is very time-consuming if some good method is not used.

The benchmark problem is summarized as follows:

- To identify the boundary models of misfiring and knocking, measurable signals such as the crank angle, engine speed, airflow rate through the throttle valve, intake pressure, and A/F are allowed to be used.

The variables in the ECU are also allowed to be used. The identified models must be simple regardless of functions of program codes.

- The identified models must cover all possible engine operation conditions, that is,

  - Engine speed: From 600 to 3,000 rpm.

  - Intake pressure: From 20 to 90 kPa (alternative of the throttle opening angle). The opening–closing speed of the throttle valve is less than 150 ms/90°.

  - Intake valve timing: For 0° to 60° of crank angle.

  - EGR valve lift: From 0 to 10 mm.

- Design a simulation to obtain sufficient data.

- Representation of the boundary condition.

- The obtained model must be evaluated on the static and transient conditions.

Benchmark Problem 2: Design of Near-Boundary Control. Near-boundary control (NBC) is defined as follows:

$$
u = \arg \min_{u \in U} \sum_{k=1}^{N} [f_{i1}(k) + f_{i2}(k) + f_{i3}(k) + f_{i4}(k)] \, dt
$$
subject to
$$
h(x, u) < 0 \tag{8.21}
$$

where $f_{ij}(k)$ is the amount of fuel injection at the engine cycle k and $j = 1, 2, 3, 4$ are the cylinder numbers. Thus, (8.21) means that the fuel consumption is minimized subject to the constraint $h(x, u) < 0$; that is, misfiring and knocking must not happen.

The designed control must be put into the controller block, and the challengers must not change the inputs and outputs from the original engine simulator. Figure 8.30 shows the inputs and outputs of the ECU block. The ECU has five inputs and four outputs.

Benchmark problem 2 is summarized as follows:

- Design the engine control to minimize fuel consumption subject to the obtained boundary models. That is, misfiring and knocking must not happen.

- The control must cover all possible engine conditions—the same as in benchmark problem 1.

- The test mode will be provided at the end of 2015.

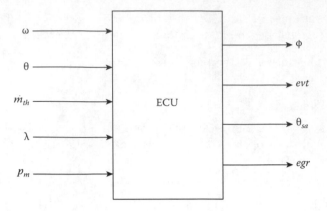

FIGURE 8.30
Inputs and outputs of ECU.

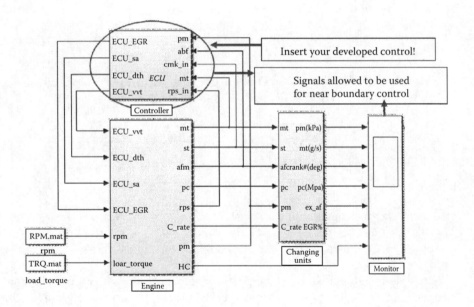

FIGURE 8.31
Top layer of the engine simulator.

Figure 8.31 shows the top layer of the engine model. The challengers can access any portion of Simulink blocks and data. However, they must not use the signals, except the signals fed to the monitor, and must put the designed controls into the controller block without any modification of the inputs and outputs. The challengers use the signals that are monitored by the monitor block. But, the others must not be used in the boundary identifications, although the challengers can access any blocks of the engine simulator.

8.4.3 Plan to Share the Results of the Challengers

Results of the challengers for this benchmark problem are expected to be shared at the Society of Automotive Engineers Japan (JSAE) Annual Congress or the Society of Instrument and Control Engineers (SICE) Annual Conference. Twenty challengers had downloaded this benchmark problem by the end of October 2014.

8.4.4 Summary

"Benchmark Problem for Boundary Modeling and Near Boundary Control" was developed in 2014. The challengers are asked to identify the misfiring and knocking boundaries of the simulator and design a control to minimize the fuel consumption subject to the constraints that misfiring and knocking must not happen on the given operation mode. The challengers are asked to put their designed controllers into the controller block, of which the inputs and outputs must not be changed by the challengers. However, the challengers can access any portions of the simulators and look into any signals by their customizations. We hope this benchmark problem developed according to strong requirements in the automotive industry contributes to promoting collaborated research between academia and the automotive industry.

Appendix A

Lyapunov Stability and Adaptive Control

In the main text of this book, the terminology of Lyapunov stability frequently appears in the design for various controllers of engines, especially in Chapter 4, where the adaptive controller of the air–fuel ratio (A/F) regulation is designed on the basis of the Lyapunov stability theory. Accordingly, this appendix collects some of the most important definitions and results of the Lyapunov stability theory, as well as the Lyapunov-based adaptive control approach, to give the reader some clues on recalling these subjects. As for the details, readers with an interest can refer to the available standard text books, such as [31, 114, 115].

A.1 Stability Definitions

Consider the system

$$\dot{x} = f(x, t), \quad x(t_0) = x_0 \tag{A.1}$$

where $x \in R^n$ and $f : R^n \times R_+ \to R^n$ are piecewise continuous in t and locally Lipschitz in x. And assume the origin is an equilibrium for (A.1), that is, $f(0, t) = 0$.

The solution of (A.1), which starts from the initial point x_0 at time $t_0 \geq 0$, is defined as $x(t; x_0, t_0)$, with $x(t_0, x_0, t_0) = x_0$, abbreviated to $x(t)$. Lyapunov stability concepts describe continuity properties of $x(t; x_0, t_0)$ with respect to x_0.

Definition A.1 Locally Lipschitz   The function f is said to be locally Lipschitz continuous in x if for some $h > 0$ there exists $l > 0$ such that

$$\|f(x_1, t) - f(x_2, t)\| \leq l\|x_1 - x_2\| \tag{A.2}$$

for all $x_1, x_2 \in B_h = \{x : \|x\| < h\}$, $t \geq 0$. The constant l is called the Lipschitz constant.   □

Definition A.2 Lyapunov Stability   The equilibrium at the origin for the system (A.1) is called stable in the sense of Lyapunov if for all $t_0 \geq 0$ and a given $\varepsilon > 0$ there exists a $\delta(\varepsilon, t_0) > 0$ such that $\|x(t; x_0, t_0)\| < \varepsilon$ for all $t \geq t_0$ whenever $\|x_0\| < \delta(\varepsilon, t_0)$.   □

Definition A.3 Uniform Stability   The equilibrium at the origin for the system (A.1) is called uniformly stable if in the preceding definition δ can be chosen independent of t_0.   □

Geometrically, this definition means that if the system starts within a distance of δ from the origin, then the system trajectory always remains within some larger, but uniformly bounded, distance ε from the origin, and δ depends on the chosen bound ε.

Definition A.4 Asymptotic Stability   The equilibrium at the origin for the system (A.1) is asymptotically stable if it is stable in the sense of Lyapunov and if it is attractive, that is, $\lim_{t \to \infty} x(t; x_0, t_0) = 0$.   □

Definition A.5 Uniform Asymptotic Stability   The equilibrium at the origin for the system (A.1) is uniformly asymptotically stable if it is uniformly stable and the trajectory $x(t; x_0, t_0)$ converges uniformly to 0; that is, there exists $\delta > 0$ and a function $\gamma(\tau, x_0) : R_+ \times R^n \to R_+$ with $\lim_{\tau \to \infty} \gamma(\tau, x_0) = 0$, $\forall x_0 \in B_\delta$ such that $\|x(t; x_0, t_0)\| < \gamma(t - t_0, x_0), \forall t \geq t_0$ whenever $\|x_0\| < \delta$.   □

The previous definitions are local, since they concern neighborhoods of the equilibrium point. Global asymptotic stability and global uniform asymptotic stability are defined as follows.

Definition A.6 Global Asymptotic Stability   The equilibrium at the origin for the system (A.1) is globally asymptotically stable if it is stable and $\lim_{t \to \infty} x(t) = 0$ for all $x_0 \in R^n$.   □

Definition A.7 Global Uniform Asymptotic Stability   The equilibrium at the origin for the system (A.1) is globally, uniformly, asymptotically stable if it is globally asymptotically stable and if in addition, the convergence to the origin of trajectories is uniform in time; that is, there is a function $\gamma : R^n \times R_+ \to R_+$ such that $\|x(t)\| < \gamma(x_0, t - t_0), \forall t \geq 0$.   □

Definition A.8 Exponential Stability   The equilibrium at the origin for the system (A.1) is exponentially stable if there exist $m > 0$, $\alpha > 0$ such that $\|x(t)\| < m e^{-\alpha(t - t_0)} \|x_0\|$ for all $x_0 \in B_h$, $t \geq t_0 \geq 0$. The constant α is called (an estimate of) the rate of convergence.   □

Global exponential stability is defined by requiring $\|x(t)\| < m e^{-\alpha(t - t_0)} \|x_0\|$ to hold for all $x_0 \in R^n$. Semiglobal exponential stability is also defined analogously except that m, α are allowed to be functions of h.

A.2   Basic Stability Theorems of Lyapunov

The so-called second method of Lyapunov, or direct method, enables one to determine the stability properties of the origin of the system (A.1) without explicitly integrating the differential equation, but instead, analyzing the properties of a generalized energy function associated with the system, termed Lyapunou function.

Definition A.9 Lyapunou Function   A scalar function $V(x, t)$ possessing the following properties is termed a Lyapunou function for the system (A.1):

1. $V(x, t)$ has continuous first-partial derivatives in x and t.

2. $V(x, t)$ is positive-definite; that is, there exists a continuous, strictly increasing scalar function $\alpha_1(\cdot)$ with $\alpha_1(0) = 0$ such that

$$V(0, t) = 0, \ V(x, t) \geq \alpha_1(\|x\|), \forall x \in B_h, t \geq 0 \tag{A.3}$$

3. The time derivative of V along the trajectory of the system (A.1) satisfies

$$\dot{V}(x, t) = \frac{\partial V}{\partial t} + \frac{\partial V}{\partial x} f(x, t) \leq 0 \tag{A.4}$$

$\square$

Theorem A.1 Local Lyapunov Stability   *The system (A.1) is locally stable in the sense of Lyapunov at the equilibrium $x = 0$ if there exists a Lyapunov function $V(x, t)$ with the properties (A.3) and (A.4).*

Theorem A.2 Local Uniform Stability   *The system (A.1) is locally uniformly stable at the equilibrium $x = 0$ if there exists a Lyapunov function $V(x, t)$ with the properties (A.3) and (A.4), and furthermore $V(x, t)$ is decrescent; that is, there exists a continuous, strictly increasing scalar function $\alpha_2(\cdot)$ with $\alpha_2(0) = 0$ such that*

$$V(x, t) \leq \alpha_2(\|x\|), \forall x \in B_h, t \geq 0 \tag{A.5}$$

Theorem A.3 Uniform Asymptotical Stability   *The system (A.1) is uniformly asymptotically stable at the equilibrium $x = 0$ if there exists $V(x, t)$ satisfying the conditions (A.3) and (A.5), and furthermore there exists a continuous strictly increasing scalar function $\alpha_3(\cdot)$ with $\alpha_3(0) = 0$ such that $\dot{V}$ along the trajectory of the system (A.1) satisfies*

$$\dot{V}(x, t) = \frac{\partial V}{\partial t} + \frac{\partial V}{\partial x} f(x, t) \leq -\alpha_3(\|x\|) \tag{A.6}$$

Theorem A.4 Global Uniform Asymptotical Stability   *The system (A.1) is globally uniformly asymptotically stable at the equilibrium $x = 0$ if there exists $V(x, t)$ satisfying the conditions (A.3) and (A.5) holding for $\forall x \in R^n$ as well as the condition (A.6), and furthermore, $\alpha_1(\|x\|) \to \infty$, and $\alpha_2(\|x\|) \to \infty$ as $\|x\| \to \infty$.*

Theorem A.5 Exponential Stability *The system (A.1) is exponentially stable at the equilibrium $x = 0$ if there exists $\gamma_i > 0, (i = 1, 2, 3)$ such that*

$$\gamma_1 \|x\|^2 \leq V(x, t)) \leq \gamma_2 \|x\|^2, \forall x \in B_h, t \geq 0 \tag{A.7}$$

$$\dot{V}(x, t) = \frac{\partial V}{\partial t} + \frac{\partial V}{\partial x} f(x, t) \leq -\gamma_3 \|x\|^2 \tag{A.8}$$

If the condition (A.7) holds for $\forall x \in R^n$, the system (A.1) is globally exponentially stable at the equilibrium $x = 0$.

A.3   Lyapunov-Based Adaptive Design

Uniform asymptotical stability is a desirable property, because systems that possess it can deal better with perturbations and disturbances. It should be noted that, in general, adaptive designs achieve less than uniform asymptotical stability. However, they achieve more than uniform stability because they force the tracking error to converge to zero. This key property is referred to as regulation when the reference signal is constant, and tracking when it is a time-varying signal [115]. For convergence analysis, a powerful tool is the following LaSalle–Yoshizawa theorem.

Theorem A.6 LaSalle–Yoshizawa *For the system (A.1), if there exist a continuously differentiable function $V(x, t)$, continuous strictly increasing scalar functions $\gamma_1(\cdot), \gamma_2(\cdot)$ with $\gamma_1(0) = 0, \gamma_2(0) = 0$, and a continuous function $W(\cdot)$ such that*

$$\gamma_1(\|x\|) \leq V(x, t)) \leq \gamma_2(\|x\|), \forall x \in R^n, t \geq 0 \tag{A.9}$$

$$\dot{V}(x, t) = \frac{\partial V}{\partial t} + \frac{\partial V}{\partial x} f(x, t) \leq -W(x) \tag{A.10}$$

then all solutions of (A.1) are globally uniformly bounded and $\lim_{t \to \infty} W(x(t)) = 0$. Furthermore, if $W(x)$ is positive-definite, the equilibrium $x = 0$ of the system (A.1) is globally uniformly asymptotically stable.

Adaptive control is a method for controlling systems with parametric uncertainty. An adaptive controller is designed by combining a parameter estimator, which provides estimates of unknown parameters, with a control law. The parameter estimator can be obtained by applying some identification methods, and the control law is designed by using the estimated parameter. Accordingly, a number of results on adaptive control design have been established. In the beginning of the 1990s, a so-called adaptive backstepping technique was proposed for the design of adaptive controllers. An important advantage of the backstepping design method is that it provides a systematic procedure to design stabilizing controllers, following a step-by-step algorithm. With this method the construction of control laws and Lyapunov functions is systematic.

The adaptive controller designed by the backstepping technique can ensure the boundedness of the closed-loop states and asymptotic tracking. To illustrate the idea of adaptive backstepping, here the following third-order strict-feedback system with linearized unknown parameters is regarded as an example.

$$\begin{cases} \dot{x}_1 = x_2 + \phi_1^T(x_1)\theta \\ \dot{x}_2 = x_3 + \phi_2^T(x_1, x_2)\theta \\ \dot{x}_3 = u + \phi_3^T(x_1, x_2, x_3)\theta \end{cases} \tag{A.11}$$

where $\theta \in R^p$ is an unknown vector constant, and $\phi_i \in R^p, i = 1, 2, 3$ are known nonlinear functions. The objective is to globally stabilize the system and also to achieve the asymptotic tracking of x_r by x_1. For the development of control laws, the following assumption is made.

Assumption A.1 The reference signal x_r and its first-, second-, and third-order derivatives are piecewise continuous and bounded.

The design procedure is elaborated in the following. Introduce the change of coordinates

$$\begin{cases} z_1 = x_1 - x_r \\ z_2 = x_2 - \alpha_1 - \dot{x}_r \\ z_3 = x_3 - \alpha_2 - \ddot{x}_r \end{cases} \tag{A.12}$$

where α_1, α_2 are called virtual controls and will be determined in later discussion.

Since θ is unknown, we can apply the idea of integrator backstepping and parameter estimation to derive the stabilizing controller with adaptive law.

Step 1. We start with the first equation of (A.11) by considering x_2 as a control variable. The derivative of tracking error z_1 is given as

$$\dot{z}_1 = z_2 + \alpha_1 + \phi_1^T\theta \tag{A.13}$$

By choosing the virtual control α_1 as

$$\alpha_1(x_1, x_r, \hat{\theta}) = -c_1 z_1 - \phi_1^T\hat{\theta} \tag{A.14}$$

where c_1 is an adjustable positive constant, $\hat{\theta}$ is an estimate of θ, the derivative of the Lyapunov function constructed as

$$V_1 = \frac{1}{2}z_1^2 \tag{A.15}$$

is rendered to satisfy the following form:

$$\dot{V}_1 = -c_1 z_1^2 + z_1 z_2 + z_1 \phi_1^T(\theta - \hat{\theta}) \tag{A.16}$$

Step 2. By the second equation of (A.11), we derive the derivative of tracking error z_2

$$\dot{z}_2 = z_3 + \alpha_2 + \phi_2^T \theta - \frac{\partial \alpha_1}{\partial x_1}(x_2 + \phi_1^T \theta) - \frac{\partial \alpha_1}{\partial x_r}\dot{x}_r - \frac{\partial \alpha_1}{\partial \hat{\theta}}\dot{\hat{\theta}} \qquad (A.17)$$

By choosing the virtual control α_2 as

$$\alpha_2(x_1, x_2, x_r, \dot{x}_r, \hat{\theta}) = -z_1 - c_2 z_2 - \phi_2^T \hat{\theta} + \frac{\partial \alpha_1}{\partial x_1}(x_2 + \phi_1^T \hat{\theta}) + \frac{\partial \alpha_1}{\partial x_r}\dot{x}_r$$

$$+ \tau_1(x_1, x_2, x_r, \dot{x}_r, \hat{\theta}) \qquad (A.18)$$

where c_2 is an adjustable positive constant and τ_1 is called the tuning function and is determined later, the derivative of the Lyapunov function recursively constructed as

$$V_2 = V_1 + \frac{1}{2}z_2^2 \qquad (A.19)$$

is rendered to satisfy the following form:

$$\dot{V}_2 = -c_1 z_1^2 - c_2 z_2^2 + z_2 z_3 + \left(z_1 \phi_1^T + z_2 \phi_2^T - z_2 \frac{\partial \alpha_1}{\partial x_1}\phi_1^T \right)(\theta - \hat{\theta})$$

$$+ z_2 \left(\tau_1 - \frac{\partial \alpha_1}{\partial \hat{\theta}}\dot{\hat{\theta}} \right) \qquad (A.20)$$

Step 3. Consider the derivative of tracking error z_3:

$$\dot{z}_3 = u + \phi_3^T \theta - \dot{\alpha}_2 - \dddot{x}_r$$

$$= u + \phi_3^T \theta - \frac{\partial \alpha_2}{\partial x_1}(x_2 + \phi_1^T \theta) - \frac{\partial \alpha_2}{\partial x_2}(x_3 + \phi_2^T \theta) - \frac{\partial \alpha_2}{\partial x_r}\dot{x}_r$$

$$- \frac{\partial \alpha_2}{\partial \dot{x}_r}\ddot{x}_r - \frac{\partial \alpha_2}{\partial \hat{\theta}}\dot{\hat{\theta}} - \dddot{x}_r \qquad (A.21)$$

By choosing the control input

$$u = -z_2 - c_3 z_3 - \phi_3^T \hat{\theta} + \frac{\partial \alpha_2}{\partial x_1}(x_2 + \phi_1^T \hat{\theta}) + \frac{\partial \alpha_2}{\partial x_2}(x_3 + \phi_2^T \hat{\theta})$$

$$+ \frac{\partial \alpha_2}{\partial x_r}\dot{x}_r + \frac{\partial \alpha_2}{\partial \dot{x}_r}\ddot{x}_r + \dddot{x}_r + \tau_2 \qquad (A.22)$$

where c_3 is an adjustable positive constant and τ_2 is also a tuning function determined later, the derivative of the Lyapunov function recursively constructed as

$$V_3 = V_2 + \frac{1}{2}z_3^2 \tag{A.23}$$

is rendered to satisfy the following form:

$$\dot{V}_3 = -c_1 z_1^2 - c_2 z_2^2 - c_3 z_3^2 + z_2 \left(\tau_1 - \frac{\partial \alpha_1}{\partial \hat{\theta}} \dot{\hat{\theta}} \right) + z_3 \left(\tau_2 - \frac{\partial \alpha_2}{\partial \hat{\theta}} \dot{\hat{\theta}} \right)$$
$$+ \left(z_1 \phi_1^T + z_2 \phi_2^T - z_2 \frac{\partial \alpha_1}{\partial x_1} \phi_1^T + z_3 \phi_3^T - z_3 \frac{\partial \alpha_2}{\partial x_1} \phi_1^T - z_3 \frac{\partial \alpha_2}{\partial x_2} \phi_2^T \right) (\theta - \hat{\theta}) \tag{A.24}$$

Step 4. Construct the Lyapunov function for the whole closed-loop system as follows:

$$V = V_3 + \frac{1}{2}(\theta - \hat{\theta})^T \Gamma (\theta - \hat{\theta}) \tag{A.25}$$

where Γ is a positive-definite matrix. Then, the derivative of V is calculated as

$$\dot{V} = \dot{V}_3 - (\theta - \hat{\theta})^T \Gamma \dot{\hat{\theta}} \tag{A.26}$$

Designing the adaptive update law as

$$\dot{\hat{\theta}} = \Gamma^{-1} \left(z_1 \phi_1 + z_2 \phi_2 - z_2 \frac{\partial \alpha_1}{\partial x_1} \phi_1 + z_3 \phi_3 - z_3 \frac{\partial \alpha_2}{\partial x_1} \phi_1 - z_3 \frac{\partial \alpha_2}{\partial x_2} \phi_2 \right) \tag{A.27}$$

and the tuning functions τ_1, τ_2 as

$$\begin{cases} \tau_1 = \dfrac{\partial \alpha_1}{\partial \hat{\theta}} \Gamma^{-1} \left(z_1 \phi_1 + z_2 \phi_2 - z_2 \dfrac{\partial \alpha_1}{\partial x_1} \phi_1 \right) \\ \tau_2 = \dfrac{\partial \alpha_2}{\partial \hat{\theta}} \dot{\hat{\theta}} + \dfrac{\partial \alpha_1}{\partial \hat{\theta}} \Gamma^{-1} z_2 \left(\phi_3 - \dfrac{\partial \alpha_2}{\partial x_1} \phi_1 - \dfrac{\partial \alpha_2}{\partial x_2} \phi_2 \right) \end{cases} \tag{A.28}$$

renders the derivative of V semi-negative-definite:

$$\dot{V} = -c_1 z_1^2 - c_2 z_2^2 - c_3 z_3^2 \tag{A.29}$$

According to the LaSalle–Yoshizawa theorem, it follows from (A.25) and (A.29) that all signals of the whole closed-loop system consisting of the system (A.11) and the controller (A.22) with the adaptive law (A.27) and tuning functions (A.28) are globally uniformly bounded and the asymptotic tracking $x_1(t) - x_r(t) \to 0$ is achieved.

Appendix B

Time-Delay System and Stability

B.1 Time-Delay System and Stability

This appendix gives a brief introduction for the stability of a time-delay system. More details for the results are referred in [28, 116].

The state-space description of a time delay has the following form:

$$\dot{x} = f(x(t - \tau)) \tag{B.1}$$

Here $x \in R^n$ denotes the vector of the n-dimensional state, τ denotes the delay time, which is smaller than a constant r, and $x(t - \tau)$ is the function in the space $C_r := \{\phi \mid \phi : [0, r] \to R^n\}$; in other words, it represents a continuous differentiable function $[0, r] \to R^n$ with respect to τ defined in the interval $[0, r]$. Depending on this function, the right-hand side of the differential equation (B.1) with time delay f is a functional, that is, $f : C_r \to R^n$. Then, this differential equation becomes a functional differential equation. For simplicity, in the following, a function $x(t - \tau)$ in the space C_r is denoted as $x_t(\tau)$ or x_t.

Therefore, the boundedness and convergence to zero of the solution $x(t)$ of the above system is guaranteed by the boundedness and convergency to zero (zero function) of the solution x_t of the following functional differential equation (B.2) in the space C_r:

$$\dot{x}(t) = f(x_t) \tag{B.2}$$

Moreover, the following norm is used to discuss the boundedness and convergency:

$$\|\phi\|_c := \sup_{0 \le \tau \le r} \|\phi(\tau)\| \tag{B.3}$$

and in the following, suppose that the functional $f : C_r \to R^n$ is Lipschitz continuous and $f(0) = 0$.

Definition B.1 For system (B.2), the solution $x = 0$ is said to be stable, if for any $\varepsilon > 0$, there exists a $\sigma(\varepsilon) > 0$ such that for any initial condition ϕ satisfying $\|\phi\|_c \le \delta$, the solution x_t satisfies

$$\|x_t(\phi)\|_c \le \varepsilon, \quad t \ge 0 \tag{B.4}$$

In addition, $x = 0$ is asymptotically stable if it is stable and as $t \to \infty$,

$$x_t(\phi) \to 0 \tag{B.5}$$

$\square$

This stability definition is the generation of the definition with respect to the differential equation, and the following theorem gives a sufficient condition for the stability.

Theorem B.1 Lyapunov–Krasovskii Stability Theorem   *For system (B.2), the solution $x = 0$ is stable if there exists a functional $V(x_t)$ that satisfies the following conditions:*

   1.   *There exist continuous nondecreasing functions $W_i(v)$, $v \geq 0$ with $W_i(0) = 0$ and $W_i(v) > 0$ for $v > 0$ ($i = 1,2$) such that*

$$W_1(\|x_t(0)\|) \leq V(x_t) \leq W_2(\| x_t(\phi) \|_c), \qquad \forall x_t \in C_r \qquad \text{(B.6)}$$

   2.   *Along all the solutions, the time derivative of $V(x_t)$ satisfies*

$$\dot{V}(x_t(\phi)) \leq 0, \qquad t \geq 0 \qquad \text{(B.7)}$$

In addition, the solution $x = 0$ is asymptotically stable if it is stable and there exists a continuous nondecreasing function W_3 such that

$$\dot{V}(x_t(\phi)) \leq -W_3(\|x_t(0)\|) \qquad \text{(B.8)}$$

where

$$\dot{V}(x_t(\phi)) := \limsup_{h \to 0^+} \frac{V(x_{t+h}(\phi)) - V(x_t(\phi))}{h} \qquad \text{(B.9)}$$

The following shows the LaSalle invariant set principle for evaluating the stability of the autonomous time-delay system (B.2).

Theorem B.2 LaSalle Invariant Set Principle   *If there exists a continuous functional $V : C \to R$ and a closed set $G \subseteq C$ relative to Equation B.2 such that*

$$\dot{V}(\phi) \leq 0, \quad \forall \Phi \in G \qquad \text{(B.10)}$$

and for any $\phi \in G$, the solution $x_t(\phi)$ is a bounded solution of Equation B.2 that remains in G, then

$$x_t(\phi) \to M \quad \text{as} \quad t \to \infty \qquad \text{(B.11)}$$

where M is the largest invariant set in E relative to Equation B.2 and the set E is defined as

$$E = \left\{ \phi \in G : \dot{V}(\phi) = 0 \right\} \qquad \text{(B.12)}$$

Appendix C

Optimal Control and Pontryagin Maximum Principle

Optimal control is an important topic in the modern control theory since it provides a systematic approach to deduce the best solutions of the control problem by optimizing some cost criterion. In essence, as an important branch of mathematics, the theoretical origin of optimal control is derived from the calculus of variations. However, it became an independent field in the 1950s thanks to the emerging development of some well-known optimization theories, such as the Pontryagin maximum principle, dynamic programming, linear quadratic control, and receding horizon control. In this book, the optimization algorithm for the receding horizon optimal problem is mainly based on the calculus of variations and Pontryagin maximum principle. We collect some fundamental knowledge about optimal control to make this book more accessible to the readers with less theoretical background.

C.1 Fundamental Knowledge for Variational Calculus

The optimization problem to be discussed involves finding the extremum values of the *functionals*. In fact, it relates to the knowledge of variational calculus. We start from several definitions.

Definition C.1 Functionals A functional is a correspondence that assigns a definite real number to each function belonging to some class. That is, a functional is a class of function where the independent variable is itself a function. A general form of a functional is given as follows:

$$J(x) = \int_{t_0}^{t_f} F\big(x(t), \dot{x}(t), t\big) dt \tag{C.1}$$

where $F(x, \dot{x}, t)$ is usually regarded as a continuous function with respect to the three arguments and x is an arbitrary continuously differentiable function defined on the closed interval $[t_0, t_f]$. $J(x)$ is a functional on the set of all such functions $x(t)$. □

To facilitate the discussion in the following, we denote $C[t_0, t_f]$ as a space that consists of all continuous functions $x(t)$ defined on $[t_0, t_f]$; while in the analogical way, the space $C^1[t_0, t_f]$ is denoted as that which involves all continuous functions $x(t)$ defined on $[t_0, t_f]$ with a continuous first derivative. With such definitions, we can talk about the *closeness* between two functions. Usually, a concept of *distance* is introduced (in many textbooks, this is replaced by *norm* for the generalized distance). For example, a distance between functions $x(t)$ and $x_0(t)$ is in general defined in the following form, in the space $C[t_0, t_f]$:

$$||x(t) - x_0(t)|| = \max_{t \in [t_0, t_f]} |x(t) - x(t_0)| \tag{C.2}$$

or in the space $C^1[t_0, t_f]$:

$$||x(t) - x_0(t)|| = \max_{t \in [t_0, t_f]} \{|x(t) - x(t_0)|, |\dot{x}(t) - \dot{x}(t_0)|\} \tag{C.3}$$

Then, we can introduce the definition for extremum values of functionals based on the above discussion.

Definition C.2 Extremum Values of Functionals   A functional $J(x(t))$ is said to get the local minimum value at the curve $x_0(t)$ if the functional value $J(x(t))$ for any curve $x(t)$ that is closed to the curve $x_0(t)$ is not less than $J(x_0(t))$, that is, $J(x(t)) - J(x_0(t)) \geq 0$.   □

Note that the closeness between the functions $x(t)$ and $x_0(t)$ can follow the definition given in (C.2) or (C.3). If we adopt the former expression, in the neighborhood of $x_0(t)$, that is,

$$||x(t) - x_0(t)|| \leq \epsilon, \ (\epsilon > 0)$$

there always exists $J(x_0) \leq J(x)$; $J(x)$ is referenced to get a relatively strict local minimum value at function $x_0(t)$.

Now, let us derive the extremum conditions of functionals. In fact, there are many similarities in the derivation in contrast to that of a real function. A derivative of a functional is introduced referring to the derivative of a real function. While before that, some definitions, such as the continuity of functionals and linear functionals, must be proposed.

Definition C.3 Continuity of Functionals   A functional $J(x(t))$ is said to be continuous at point $x_0(t)$ if for any given positive ϵ, there exists a positive δ such that

$$|J(x(t)) - J(x_0(t))| < \epsilon$$

for all $x(t)$ such that $||x(t) - x_0(t)|| < \delta$.   □

Definition C.4 Linear Functionals   A continuous functional $J(x(t))$ is linear if it satisfies the following two conditions:

$$J(x_1(t), x_2(t)) = J(x_1(t)) + J(x_2(t))$$
$$J(\gamma x(t)) = \gamma J(x(t))$$

where γ is an arbitrary constant.   □

Recall the definition of the derivative of a real function as follows. For a continuously differential function $x = f(t)$, its incremental form can be represented by

$$\Delta x = f(t + \Delta t) - f(t) = f'(t)\Delta t + \eta(t, \Delta t)$$

In the above expression, the last item is a nonlinear function of Δt, and it can be ignored when Δt takes infinitesimal. Thus, the first item of the right side plays a main role in influencing the Δx, and it is called a derivative of the function, denoted by

$$dx = f'(t)dt$$

Similarly, the above derivation process can be further applied to the derivative of functionals. We can now give the definition in the following.

Definition C.5 Derivative of Functionals   For a continuous functional $J(x(t))$, if its incremental form can be given by

$$\begin{aligned}\Delta J &= J(x(t) + \delta x(t)) - J(x(t)) \\ &= \dot{J}(x(t), \delta x(t)) + \eta(x(t), \delta x(t))\end{aligned} \tag{C.4}$$

where the second item $\eta(\cdot) \to 0$ as $\delta x(t) \to 0$ and the first item $\dot{J}(\cdot)$ is a linear functional with respect to $\delta x(t)$, then $\dot{J}(x(t), \delta x(t))$ is called the derivative of functional $J(x(t))$, denoted by

$$\delta J(x(t)) = \dot{J}(x(t), \delta x(t)) \tag{C.5}$$

Besides, it can be said $J(x(t))$ is differentiable at point x if there exists the derivative or its incremental form can be written in the form given in (C.4).   □

It is well known in elementary mathematics that a necessary condition for a differentiable function $x = f(t)$ to have a local extremum at t_0 is that $f'(t_0) = 0$. A similar necessary condition for a differentiable functional $J(x(t))$ can be derived. Here we only give some important results without the proof.

Theorem C.1   *The derivative of a differentiable functional $J(x(t))$ at a point $x(t)$ is unique.*

Theorem C.2   *If a differentiable functional $J(x(t))$ has a local extremum value at $x_0(t)$, then $\delta J(x_0(t)) = 0$.*

Now let us discuss the basic variational calculus problem and Euler–Lagrange equation. Assume $x(t)$ is at least a first differentiable function, that is, $x(t) \in C^1[t_0, t_f]$, and $F(x, \dot{x}, t)$ is a function with continuous first and second partial derivatives with respect to $(x, \dot{x}, t)$. Then a variational problem is to find a $x(t)$ to be an extremum of the functional

$$J(x) = \int_{t_0}^{t_f} F\big(x(t), \dot{x}(t), t\big) dt \tag{C.6}$$

such that the boundary conditions $x(t_0) = x_{t_0}$ and $x(t_f) = x_{t_f}$. In fact, the Euler–Lagrange equation provides a necessary condition to solve this kind of problem.

Theorem C.3 Euler–Lagrange Equation *If a functional given by (C.6) has an extremum at $x_0(t)$, then $x_0(t)$ satisfies the Euler–Lagrange equation:*

$$\frac{\partial F}{\partial x}\big(x_0(t), \dot{x}_0(t), t\big) = \frac{d}{dt}\Big(\frac{\partial F}{\partial x'}\big(x_0(t), \dot{x}_0(t), t\big)\Big), \ \ t \in [t_0, t_f] \tag{C.7}$$

Note that the above theorem implies a general solution for the variational problem with the curve $x_0(t)$, which has the fixed two-point boundary value. Moreover, if the boundary conditions are not explicitly specified, that is, we will tackle the variational problem with free boundary conditions, then the extremum condition at $x_0(t)$ should have some additional transversal conditions added:

$$\frac{\partial F}{\partial x'}\big(x_0(t), \dot{x}_0(t), t\big)|_{t=t_0} = 0 \ \text{and} \ \frac{\partial F}{\partial x'}\big(x_0(t), \dot{x}_0(t), t\big)|_{t=t_f} = 0 \tag{C.8}$$

except the Euler–Lagrange equation (C.7).

Besides, in many cases, the constrained conditions are not only on the boundary point, but also on some middle points or on the whole trajectory for $x(t)$ (named *path constraints*). For such problems, we have the following result.

Theorem C.4 *If $x(t) \in C^1[t_0, t_f]$ enables the functional whose form is given by (C.6) to get the extremum such that the following equality conditions apply*

$$\Psi(x(t), \dot{x}(t), t) = 0, \ \text{for all } t \in [t_0, t_f] \tag{C.9}$$

then there must exist some proper functions $\lambda(t)$ that make the functional

$$J(x) = \int_{t_0}^{t_f} \big[F\big(x(t), \dot{x}(t), t\big) + \lambda(t)\Psi\big(x(t), \dot{x}(t), t\big)\big] dt \tag{C.10}$$

obtain the extremum value at $x(t)$, where $x(t)$ satisfies the Euler–Lagrange equation:

$$\frac{\partial H_0}{\partial x}\big(x(t), \dot{x}(t), t\big) = \frac{d}{dt}\Big(\frac{\partial H_0}{\partial x'}\big(x(t), \dot{x}(t), t\big)\Big), \ \ t \in [t_0, t_f] \tag{C.11}$$

and

$$H_0(x(t), \dot{x}(t), \lambda(t), t) = F\big(x(t), \dot{x}(t), t\big) + \lambda(t)\Psi(x(t), \dot{x}(t), t) \tag{C.12}$$

So far, we have discussed the definition of the functional and a necessary condition (Euler–Lagrange equation) for the extremum of a functional. In essence, the optimal control problem is just based on the above concepts, and it is to find the extremum of a functional with a specific control action. We will summarize the basic form of the optimal control problem in the following section.

C.2   Optimal Control Problem

Consider a continuous function u defined on an interval $[t_0, t_f]$, and denote $F(x, u, t)$ and $f(x, u)$ to be continuously differentiable functions with respect to their arguments. Besides, (x, u) have the following relations in $t \in [t_0, t_f]$:

$$\dot{x}(t) = f(x(t), u(t)), \ x(t_0) = x_{t_0} \tag{C.13}$$

and a functional $J(u)$ is given as

$$J(u) = \int_{t_0}^{t_f} F(x(t), u(t), t) dt \tag{C.14}$$

Then, the typical optimal control problem can be formulated to find a control trajectory $u \in C[t_0, t_f]$ to minimize the functional (C.14) such that the constrained conditions (C.13) are met.

This is a mathematical description that is analogous to the aforementioned variational problem. Indeed, a practical system dynamics can be in general described by (C.13) with a known initial condition $x(t_0) = x_{t_0}$ at time t_0. If a function $u^*(t)$ is a solution of such a problem, then $u^*(t)$ is referred to as an *optimal control input*. Under the optimal control input, the system dynamical state $x^*(t)$ can be derived based on (C.13), which is called the *optimal state*. The optimized functional is said to be a *performance index*.

We now consider the extremum conditions for the proposed optimal control problem. First, an assumption will be held in the following discussion that the function $u \in C[t_0, t_f]$ has no constraints on itself. Then, the following results can be easily derived based on Theorem C.4.

Theorem C.5 *For the above optimal control problem, if $u^*(t)$ is an optimal control input and accordingly $x^*(t)$ is the optimal state, then there exists a $\lambda^*(t) \in C^1[t_0, t_f]$ such that*

$$\frac{\partial F}{\partial x}(x^*(t), \dot{x}^*(t), t) + \lambda^*(t)\frac{\partial f}{\partial x}(x^*(t), u^*(t)) = -\dot{\lambda}^*(t), \ t \in [t_0, t_f], \lambda^*(t_f) = 0 \tag{C.15}$$

$$\frac{\partial F}{\partial u}(x^*(t), \dot{x}^*(t), t) + \lambda^*(t)\frac{\partial f}{\partial u}(x^*(t), u^*(t)) = 0, \ t \in [t_0, t_f] \tag{C.16}$$

It should noted that Theorem C.5 just gives a necessary condition for optimality and (C.16) implies $u(t)$ should not be constrained.

C.3   Hamiltonian and Pontryagin Maximum Principle

In a practical control system, it is well known that the manipulated range of the actuator always exists in the boundary conditions. In other words, the control input $u(t)$ is actually with some constraints. Thus, the above optimality conditions (Theorem C.5) cannot be applied to such an optimal control problem. To further discuss the optimality conditions for such a constrained optimization problem, we introduce the concept of Hamiltonian, which is able to make the discussion of the optimality conditions in a unified framework.

In general, Hamiltonian is defined in the following form:

$$H(x(t), u(t), \lambda(t), t) = F(x(t), u(t), t) + \lambda(t) f(x(t), u(t)) \qquad (C.17)$$

Then, with this definition, it is obvious that the conditions (C.15) and (C.16) given in Theorem C.5 can be equivalently converted to the following form:

$$\frac{\partial H}{\partial x}(x^*(t), u^*(t), \lambda^*(t), t) = -\dot{\lambda}^*(t), \ t \in [t_0, t_f], \ \lambda^*(t_f) = 0 \quad (C.18)$$

$$\frac{\partial H}{\partial u}(x^*(t), u^*(t), \lambda^*(t), t) = 0, \ t \in [t_0, t_f] \qquad (C.19)$$

Besides, we can find a similar form for differential equation $\dot{x} = f(x(t), u(t))$, that is,

$$\frac{\partial H}{\partial \lambda}(x^*(t), u^*(t), \lambda^*(t), t) = \dot{x}^*(t), \ t \in [t_0, t_f], \ x^*(t_0) = x_{t_0} \qquad (C.20)$$

In such statements, the function $\lambda^*(t)$ is referred to as the *co-state* and Equation C.18 is called the *adjoint differential equation*. In addition, (C.18) and (C.20) comprise a *Hamiltonian differential system*. Now, by using the same framework, we can introduce the optimality condition for the constrained optimization problem.

Theorem C.6 Pontryagin Minimum Principle   *Suppose $F(x, u, t)$ and $f(x, u)$ are the continuously differentiable functions corresponding to their arguments and $u \in C[t_0, t_f]$ is with constraints on itself, such as u belong to a set U that consists of all such admissible functions. If $u^*(t)$ is an optimal control input for*

$$J(u) = \int_{t_0}^{t_f} F(x(t), u(t), t) dt \qquad (C.21)$$

subject to the differential equations

$$\dot{x}(t) = f(x(t), u(t)), \ t \in [t_0, t_f], \ x(t_0) = x_{t_0} \tag{C.22}$$

and

$$u(t) \in U \tag{C.23}$$

and if $x^*(t)$ is the corresponding optimal state, then there exists a $\lambda^*(t) \in C^1[t_0, t_f]$ such that

$$\frac{\partial H}{\partial x}(x^*(t), u^*(t), \lambda^*(t), t) = -\dot{\lambda}^*(t), \ t \in [t_0, t_f], \ \lambda^*(t_f) = 0 \tag{C.24}$$

$$\frac{\partial H}{\partial \lambda}(x^*(t), u^*(t), \lambda^*(t), t) = \dot{x}^*(t), \ t \in [t_0, t_f], \ x^*(t_0) = x_{t_0} \tag{C.25}$$

for all $t \in [t_0, t_f]$

$$H(x^*(t), u^*(t), \lambda^*(t), t) = \min_{u(t) \in U} H(x^*(t), u(t), \lambda^*(t), t) \tag{C.26}$$

holds.

The Pontryagin minimum principle also provides a necessary condition for optimality. In fact, it does not require the differentiability of function u. Hence, it can be applied in a wide scope in the practical control system, and indeed, Theorem C.5 is a corollary of the Pontryagin minimum principle.

In addition, we only focus on the integral form of the performance index in the above discussion. If a performance index is given with a terminal cost in the following form,

$$J(u) = \Phi(x(t_f), t_f) + \int_{t_0}^{t_f} F(x(t), u(t), t) dt \tag{C.27}$$

then for the above constrained optimization problem, the same results can be derived by changing the expression of terminal conditions in the adjoint differential equation (C.24), that is,

$$\frac{\partial H}{\partial x}(x^*(t), u^*(t), \lambda^*(t), t) = -\dot{\lambda}^*(t), \ t \in [t_0, t_f] \tag{C.28}$$

$$\lambda^*(t_f) = \frac{\partial \Phi(x(t_f), t_f)}{\partial x(t_f)} \tag{C.29}$$

The proof is omitted in this appendix. Interested readers can refer to textbooks such as [117–119].

Appendix D

Stochastic Optimal Control

In this appendix, we collect several definitions and basic results from stochastic system control theory. The goal is to provide some basic terminology and concepts related to the stochastic control design technique used in this book. For detailed presentations, readers with an interest can refer to textbooks such as [120–125].

D.1 Markov Process

This section introduces in a simple way the skeletal structure of the basic mathematical concepts related to the Markov process [120, 122].

Definition D.1 Event An event is the occurrence or nonoccurrence of a phenomenon.

The set of elementary events (samples or possible outcomes of an experiment) will be denoted by $\Omega = \{\omega\}$. □

Definition D.2 σ-Algebra The system $\mathcal{F}$ of subsets of Ω is said to be the σ-algebra associated with Ω, if the following properties are fulfilled: □

1. $\Omega \in \mathcal{F}$.

2. For any set $A_n \in \mathcal{F}$ $(n = 1, 2, \ldots)$ the countable union of elements in $\mathcal{F}$ belongs to the σ-algebra $\mathcal{F}$, as well as the intersection of elements in $\mathcal{F}$

$$\bigcup_{n=1}^{\infty} A_n \in \mathcal{F}, \quad \bigcap_{n=1}^{\infty} A_n \in \mathcal{F} \tag{D.1}$$

3. For any set $A \in \mathcal{F}$, its complement belongs to the σ-algebra: $\bar{A} = \{\omega \in \Omega | \omega \notin A\} \in \mathcal{F}$.

In other words, the σ-algebra is a collection of subsets of the set Ω of all possible outcomes of an experiment, including the empty set $\emptyset$. Simply put, the σ-algebra represents a collection of any possible events.

Definition D.3 Probability Measure The probability measure (i.e., a function mapping $\mathcal{F} \to \mathcal{R}^{\infty}$) that assigns a probability to each set in the field $\mathcal{F}$ will be denoted by P. This function satisfies the following conditions:

1. For ever $A \in \mathcal{F}$, $P(A) \geq 0$, $P(\Omega) = 1$.
2. For every mutually disjoint set $A_1, A_2, \ldots$ in $\mathcal{F}$,

$$P \left(\bigcup_{k=1}^{\infty} A_k \right) = \sum_{k=1}^{\infty} P(A_k), \quad A_i \bigcap A_j = \emptyset, (i \neq j) \tag{D.2}$$

Definition D.4 Probability Space   The triple $(\Omega, \mathcal{F}, P)$ is called a probability space. A random variable is a real function defined over a probability space, assuming real values. In other words, a random variable is a quantity that is measured in connection with a random experiment.   □

Definition D.5 Measurable Random Variable   We say that $X(\omega)$ is measurable with respect to a sigma field $\mathcal{F}$ of subsets of Ω, or more briefly, $\mathcal{F}$-measurable, if $\{\omega : X(\omega) < x\} \in \mathcal{F}$ for all real x.   □

Definition D.6 Stochastic Process   A stochastic process $\{X_t, t \in T\}$, $T \in R^1$, $X_t \in R^n$, is a family of random variables indexed by the parameter t and defined on a common probability space $(\Omega, \mathcal{F}, P)$. For each t, $\omega \in \Omega$, $X_t(\omega)$ is a random variable. For each ω, $t \in T$, $X_t(\omega)$ is called a sample function or realization of the process.   □

The stochastic process is said to be continuous if T is a continuous subset on R^1 and is said to be discrete if T is a finite or countable set from R, that is, $\{t_l, t_2, \ldots, t_n, \ldots\}$.

Definition D.7 Stationary Process   A stochastic process $x(t)$ is said to be stationary if its statistics (distribution functions for each fixed t) are not affected by a shift in time, that is, the two processes $x(t)$ and $x(t+\varepsilon)$ have the same statistics for any ε.   □

Definition D.8 Ergodic Process $x(t)$ is said to be ergodic in the most general form (with probability 1) if all its statistics can be determined from the time averages of a process instead of the ensemble averages.   □

Definition D.9 Probability Distribution   The probability distribution gives the relationship (correspondence) between all possible values (realizations) of a given random variable and their associated probabilities.   □

Certainly, a table is the most simple form for the representation of this correspondence. Let $x_i (i = 1, 2, \ldots, K)$ represent the possible values of the random variable X, and $p_i (i = 1, 2, \ldots, K)$ the corresponding probabilities. The (cumulative) probability distribution of a continuous variable X will be denoted by $F(x)$ and defined as follows: $F(x) = P(X < x)$. The distribution function contains all the information that is relevant to the probability theory. The probability distribution has the following properties:

1. $F(x)$ is a nondecreasing function.
2. $\lim_{x \to -\infty} F(x) = 0$ and $\lim_{x \to +\infty} F(x) = 1$.
3. $F(x)$ is continuous at least from the left.

Definition D.10 Probability Density Consider the probability that a continuous random variable X belongs to the interval $[x, x + \Delta x]$. This probability is obviously equal to the variation of the probability distribution in this interval, that is,

$$P(x < X \leq x + \Delta x) = F(x + \Delta x) - F(x) \tag{D.3}$$

$\square$

Assume that $F(x)$ is continuous and differentiable, and consider the mean of this probability toward the length unit. If $\Delta x \to 0$, we obtain

$$\lim_{\Delta x \to 0} \frac{F(x + \Delta x) - F(x)}{\Delta x} = \frac{dF(x)}{dx} = f(x) \tag{D.4}$$

where $f(x)$ represents the probability density function (pdf). In summary, every real valued function that is nonnegative $(P(\cdot) \geq 0)$, integrable over the whole real axis, and satisfies (D.4) is the probability density of a random variable X.

In many situations, the probability density function is a priori unknown. One possible way to obtain a mathematical expression of $f(x)$ is to consider a model of the form

$$f(x) = \sum_{i=1}^{\infty} \alpha_i \phi_i(x) \tag{D.5}$$

where $\phi_i(x)(i = 1, 2, \ldots)$ are a priori known functions. A truncated series of m terms

$$f(x) = \sum_{i=1}^{m} \alpha_i \phi_i(x) \tag{D.6}$$

can then be used as an approximation of the true function.

Definition D.11 Expectation The Lebesgue integral,

$$E\{\xi\} := \int_{x \in X} \xi(x) dF(x) \tag{D.7}$$

is said to be the mathematical expectation of a random variable ξ having a distribution function $F(x)$ given on X. This integral is understood in the Lebesgue sense. If $F(x)$ is differentiable, $f(x) = dF(x)/dx$, then

$$E\{\xi\} = \int_{x \in X} \xi(x) \frac{dF(x)}{dx} dx = \int_{x \in X} \xi(x) f(x) dx \tag{D.8}$$

In the discrete case, the expectation corresponds to the sum of the random variables weighted by the probability with which they are assumed. $\square$

Definition D.12 Variance or Second Moment   The variance represents another characteristic of a given random variable. It characterizes the fluctuation of a given random variable around its mean value.   □

The variance of a given random variable ξ is given by

$$Var\{\xi\} := E\{(\xi - E\{\xi\})^2\} = E\{(\xi)^2\} - (E\{\xi\})^2 = \sigma_\xi^2 \quad \text{(D.9)}$$

The variance of the sum of independent random variables is equal to the sum of the variances of these random variables. In fact, if $S_n = \sum_{i=1}^n \xi_i$, then

$$\sigma_{S_n}^2 = E\{(S_n - E\{S_n\})^2\} = \sum_{i=1}^n \sigma_{\xi_i}^2 \quad \text{(D.10)}$$

Definition D.13 Centering   Centering corresponds to a variable change. In other words, we say that ξ is centered at c if we replace it by $\xi - c$.   □

From the previous definition, the following properties can be deduced:

1. A random variable is centered at its expectation if and only if its expectation exists and is equal to zero.

2. The variance of a given random variable remains unchanged by centering, that is,

$$Var\{\xi\} := E\{(\xi - E\{\xi\})^2\} = E\{(\xi - c - E\{\xi - c\})^2\} = \sigma_\xi^2 \quad \text{(D.11)}$$

Definition D.14 Space L_N   Let ξ be a random variable such that $E\{|\xi|^m\}, m = 1, 2, \ldots, N$, exists and forms a space L_N over the probability space $(\Omega, \mathcal{F}, P)$. In symbols, $\xi \in L_N$. The space L_2 plays an important role in the investigation of probability problems, especially in those related to the sums of independent random variables.   □

Definition D.15 Conditional Expectation   The random variable $E\{\xi|\mathcal{F}_0\}$ is called the conditional mathematical expectation of the random variable $\xi(\omega)$ given on $(\Omega, \mathcal{F}, P)$ with respect to the σ-algebra $\mathcal{F}_0 \subseteq \mathcal{F}$ if

1. It is $\mathcal{F}_0$-measurable, that is,

$$\{\omega|\ E\{\xi|\mathcal{F}_0\} \le x\} \in \mathcal{F}_0, \ \forall x \in R^1 \quad \text{(D.12)}$$

2. For any set $A \in \mathcal{F}_0$,

$$\int_{\omega \in A} E\{\xi|\mathcal{F}_0\}P\{d\omega\} = \int_{\omega \in A} \xi(\omega)P\{d\omega\} \quad \text{(D.13)}$$

Definition D.16 Markov Property   The Markov property simply states that the present state of the system completely determines the probability for the next step into the future. In other words, we have a fundamental one-step dependence.   □

Definition D.17 Markov Chain   A finite stationary Markov chain consists of a set $X = (x(1), \cdots, x(N))$ of states and a probability transition matrix

$$\Pi = [p_{ij}], i, j = 1, 2, \ldots, N, \ \ p_{ij} \in [0,1], \ \sum_{j=1}^{N} p_{ij} = 1, (i = 1, 2, \ldots, N)$$

$$(\text{D.14})$$

where p_{ij} represents the transition probability from the state x_i to the state x_j, that is,

$$p_{ij} = P[x_{n+1} = x(j)|x_n = x(i)] \tag{D.15}$$

Here x_n represents the state occupied at time n $(n = 0, 1, \ldots)$. □

Notice that the ith row $(i = 1, 2, \ldots, N)$ of Π represents the transition probability from the state $x(i)$ to the states $x(j)(j = 1, 2, \ldots, N)$.

For finite controlled Markov chains, the probability transition matrix depends on the actions and is denoted by

$$\Pi_k = [p_{ij}^k], \ i, j = 1, 2, \ldots, N, k = 1, 2, \ldots, K \tag{D.16}$$

$$p_{ij}^k = Pr[x_{n+1} = x(j)|x_n = x(i), u_n = u(k)] \tag{D.17}$$

where the index k corresponds to the action $u_n(k) \in U = (u(1), \cdots, u(K))$ selected at time n.

Definition D.18 Markov Processes   Let t_i and t be elements of the index set T such that $t_1 < t_2 < \cdots < t_k < t$. A stochastic process $\{x(t), t \in T\}$ is called a Markov process if

$$P\{x(t) \leq \xi|x(t_1), x(t_2), \ldots, x(t_k)\} = P\{x(t) \leq \xi|x(t_k)\} \tag{D.18}$$

where $P\{\cdot|x(t_k)\}$ denotes the conditional probability given $x(t_k)$. If the probability distribution of $x(t_1)$, the initial probability distribution

$$F(\xi_1; t_1) = P\{x(t_1) \leq \xi_1\} \tag{D.19}$$

and the transition probability distribution

$$F(\xi_t, t|\xi_s, s) = P\{x(t) \leq \xi_t|x(s) = \xi_s\} \tag{D.20}$$

are given, it follows from Bayes' rule that the distribution function of the random variable $x(t_1), x(t_2), \cdots, x(t_k)$ is given by

$$\int_0^{\xi_1} \int_0^{\xi_2} \cdots \int_0^{\xi_{n-1}} F(\xi_n, t_n|\eta_{n-1}, t_{n-1}) dF(\eta_{n-1}, t_{n-1}|\eta_{n-2}, t_{n-2}) \cdots dF(\eta_1, t_1)$$

$$(\text{D.21})$$

A Markov process is thus defined by two functions: the absolute probability distribution $F(\eta; s)$ and transition probabilities $F(\xi, t|\eta, s)$. □

Definition D.19 Gauss–Markov Process   A Gauss–Markov process is simply a random process that is both Gaussian and Markov. Because of these properties—that Markov processes can be completely described by an initial condition and a transition density function, which is a great simplification, and linear transformations of Gaussian random vectors remain Gaussian—a Gauss–Markov process can be represented by the state vector of a multistate linear dynamic system:

$$x_{k+1} = \Phi_k x_k + w_k \tag{D.22}$$

where x_k is an n-vector. Φ_k is an $n \times n$ known matrix. The driving function or process noise w_k is an n-vector-valued Gaussian random sequence, whose statistics will be assumed to be

$$E\{x_k\} = \bar{w}_k, \quad E[(w_k - \bar{w}_k)(w_l - \bar{w}_l)^T] = W_k \delta_{kl} \tag{D.23}$$

with the so-called Kronecker delta $\delta_{kl} = 1(k = l)$, or $\delta_{kl} = 0(k \neq l)$. Also assume that the initial state is random and Gaussian, with the following statistics:

$$E\{x_0\} = \bar{x}_0, \quad E[(x_0 - \bar{x}_0)(x_0 - \bar{x}_0)^T] = P_0, \quad E[(x_0 - \bar{x}_0)(w_k - \bar{w}_k)^T] = 0 \tag{D.24}$$

□

D.2   Stochastic Stability

In this section, some stability definitions and theorems of the stochastic system will be presented [121, 123]. Consider the system

$$\dot{x} = f(x, t, \xi(t, \omega)), \quad x(t_0) = x_0(\omega) \tag{D.25}$$

where $x \in R^n$ and $\xi(t, \omega)(t \geq 0)$ are a separable measurable stochastic process with values in R^k. f is a Borel-measurable function satisfying the following conditions:

1. There exists a stochastic process $B(t, \omega) \in L$, namely, which is absolutely integrable over every finite interval, such that

$$\|f(x_2, t, \xi(t, \omega)) - f(x_1, t, \xi(t, \omega))\| \leq B(t, \omega)\|x_2 - x_1\|, \forall x_1, x_2 \in R^n \tag{D.26}$$

2. The process $f(0, t, \xi(t, \omega))$ is in L, that is, for every $T > 0$,

$$P\left\{ \int_0^T |f(0, t, \xi(t, \omega))| dt < \infty \right\} = 1 \tag{D.27}$$

Definition D.20 Stable in Probability The solution $x(t) = 0$ of the system (D.25) is called stable in probability if for any $\varepsilon > 0$, $\rho > 0$, there exists a $\delta(\rho, \varepsilon, t_0) > 0$ such that

$$\|x_0\| < \delta(\rho, \varepsilon, t_0) \implies P\{\|x(t, \omega, x_0, t_0)\| > \varepsilon\} < \rho \qquad \text{(D.28)}$$

$\square$

Definition D.21 Asymptotically Stable in Probability The solution $x(t) = 0$ of the system (D.25) is called asymptotically stable in probability if it is stable in probability and, for each $\varepsilon > 0$, there exists a $\delta(t_0) > 0$ such that

$$\|x_0\| < \delta \implies P\{\lim_{t \to \infty} \|x(t, \omega, x_0, t_0)\| > \varepsilon\} \to 0 \qquad \text{(D.29)}$$

$\square$

Definition D.22 Exponentially Stable in Probability The solution $x(t) = 0$ of the system (D.25) is called exponentially stable in probability if for any $\varepsilon > 0$, $\rho > 0$, there exist $\delta(\rho, \varepsilon) > 0$ and $\alpha > 0$ such that

$$\|x_0\| < \delta \implies P\{\|x(t, \omega, x_0, t_0)\| > \varepsilon\} \leq \rho e^{-\alpha(t-t_0)}, \forall t \geq t_0 > 0 \qquad \text{(D.30)}$$

$\square$

Definition D.23 Bounded in Probability The solution $x(t) = 0$ of the system (D.25) is called bounded in probability if there exists a $M > 0$ such that

$$\lim_{T \to \infty} P\{\sup_{t \geq T} \|x(t, \omega, x_0, t_0)\| \leq M\} = 1 \qquad \text{(D.31)}$$

$\square$

Definition D.24 p-Stable The solution $x(t) = 0$ of the system (D.25) is called p-stable if for each $\varepsilon > 0$, there exists a $\delta(\varepsilon, t_0) > 0$ such that

$$\|x_0\| < \delta(\varepsilon, t_0) \implies E\|x(t, \omega, x_0, t_0)\|^p < \varepsilon \qquad \text{(D.32)}$$

$\sqcup$

Definition D.25 Asymptotically p-Stable The solution $x(t) = 0$ of the system (D.25) is called asymptotically p-stable if it is p-stable and there exists a $\delta(t_0) > 0$ such that

$$\|x_0\| < \delta \implies \lim_{t \to \infty} E\|x(t, \omega, x_0, t_0)\|^p = 0 \qquad \text{(D.33)}$$

$\square$

Definition D.26 Exponentially p-Stable The solution $x(t) = 0$ of the system (D.25) is called exponentially p-stable if there exist constants $A > 0$ and $\alpha > 0$ such that

$$E\|x(t, \omega, x_0, t_0)\|^p \leq A\|x_0\|^p exp\{-\alpha(t - t_0)\} \qquad \text{(D.34)}$$

$\square$

It should be noted that for the system (D.25) the stability conditions can be given in terms of a Lyapunov function $V(x,t) > 0$ such that $E\dot{V}(x,t) < 0$; however, in order to calculate the expectation $E\dot{V}(x,t)$, one must solve the system (D.25) with a suitable initial condition. Thus, this limits the practical use of the criterion.

For linear discrete-time stochastic systems with a Markov chain,

$$x(k+1) = A(r(k))x(k) + B(r(k))w(k), \quad x(0) = x_0, r(0) = r_0 \qquad \text{(D.35)}$$

where $\{r(k)\}$ is a discrete-time Markov chain with finite state space $\{1,\ldots,N\}$ and transition probability matrix $P(r_{k+1} = j | r_k = i) = [p_{ij}]$. $\{x(k)\}$ denotes the random state sequence in R^n and $\{w(k)\}$ random disturbances in R^m.

Definition D.27 Stochastically Stable   The system (D.35) is said to be stochastically stable if for every initial (x_0, r_0), there exists a finite number $M(x_0, r_0) > 0$ such that

$$\lim_{N \to \infty} E\left\{ \sum_{k=0}^{N} x_k^T(x_0, r_0) x_k(x_0, r_0) \right\} < M(x_0, r_0) \qquad \text{(D.36)}$$

$\square$

Definition D.28 Mean Square Stable   The system (D.35) is said to be mean square stable if for any initial condition x_0 and initial distribution r_0, there exist q, Q independent of x_0 such that

$$\lim_{k \to \infty} \|q(k) - q\| \to 0, \quad \lim_{k \to \infty} \|Q(k) - Q\| \to 0 \qquad \text{(D.37)}$$

where $q(k) := E\{x(k)\}$, $Q(k) := E\{x(k)x^T(k)\}$. $\square$

Theorem D.1 Mean Square Stability   *For the system (D.35), if for any given set of symmetric matrices $\{W(s_i) > 0, i = 1,\ldots,N\}$ there exists a set of symmetric solutions $\{\chi(s_i), i = 1,\ldots,N\}$ such that*

$$\sum_{j=-1}^{N} p_{ij} A^T(s_i) \chi(s_j) A(s_i) - \chi(s_i) = -W(s_i) \qquad \text{(D.38)}$$

then the system (D.35) is mean square stable.

D.3   Stochastic Optimization

Consider a discrete-time stochastic system in terms of the difference equation [124, 125]

$$x_{k+1} = f_k(x_k, u_k, w_k), \quad k = 0, 1, \cdots \qquad \text{(D.39)}$$

where $x_k \in S, u_k \in C, w_k \in D$, S, C, D are finite sets and $S = \{1, 2, \cdots, s\}$. u_k is constrained to take values in a given nonempty subset $U(x_k)$ of C, that is, $u_k \in U(x_k), \forall x_k \in S$. The random disturbances $w_k, k = 0, 1, \cdots$ have identical statistics and are characterized by probabilities $P(\cdot|x_k, u_k)$ defined on D, where $P(\cdot|x_k, u_k)$ is the probability of occurrence of w_k, when the current state and control are x_k and u_k, respectively, that is, the probability distribution

$$p_{ij}(u, k) = P_k\{W_k(i, u, j)|x_k = i, u_k = u\}$$

with the set $W_k(i, u, j) = \{w|j = f_k(i, u, w)\}$, which may depend explicitly on x_k and u_k but not on values of prior disturbances $w_{k-1}, \cdots, w_0$.

In the following we present an optimal control problem of the stochastic system over a finite number of stages (a finite horizon) and an infinite horizon, respectively.

Stochastic Optimal Control with Finite Horizon. For the system (D.39) the optimal control problem is to find a control law (control policy) π^* in all admissible policies $\pi = \{\mu_0, \mu_1, \cdots, \mu_{N-1}\}$, where μ_k maps states x_k into controls $u_k = \mu_k(x_k)$ and $\mu_k(x_k) \in C, \forall x_k \in S, k = 0, 1, \cdots, N - 1$, to minimize the expected cost function described as follows:

$$J_\pi(x_0) = \underset{w_k}{E} \left\{ g_N(x_N) + \sum_{k=0}^{N-1} g_k(x_k, \mu_k(x_k), w_k) \right\} \qquad (D.40)$$

where the cost per stage $g : S \times C \times D \to R$ is given and bounded.

It should be noted that the expectation is with respect to the joint distribution of the random variable involved, which takes into account the case that the cost is generally a random variable and cannot be meaningfully optimized because of the presence of w_k. Also, the optimization is over the controls $u_0, u_1, \cdots, u_{N-1}$, but each control u_k is selected with some knowledge of the current state x_k (either its exact value or some other information relating to it). Accordingly, for a given initial state x_0, an optimal policy π^* is one that minimizes the above cost, that is,

$$J_{\pi^*}(x_0) = \min_{\pi \in \Pi} J_\pi(x_0)$$

where Π is the set of all admissible policies.

Dynamic programming (DP) is an appropriate approach to solve the above optimal control problem. It is worth mentioning that the dynamic programming technique rests on the following principle of optimality:

Let $\pi^* = \{\mu_0^*, \mu_1^*, \cdots, \mu_{N-1}^*\}$ be an optimal policy for the above stochastic optimal control problem, and assume that a given state x_i occurs at time i when using π^*. The truncated policy $\pi_i^* = \{\mu_i^*, \mu_{i+1}^*, \cdots, \mu_{N-1}^*\}$ is an optimal solution to the subproblem that minimizes the following function from time i to time N with x_i at time i:

$$E\left\{ g_N(x_N) + \sum_{k=i}^{N-1} g_k(x_k, \mu_k(x_k), w_k) \right\}$$

That is, if $J^*(x_0) = \min_\pi J(x_0) = J_{\pi^*}(x_0)$, then $J_i^*(x_i) = \min_{\pi_i} J_i(x_i) = J_{\pi_i^*}(x_i)$.

DP algorithm: For every initial state x_0, the optimal cost $J^*(x_0)$ of the evaluated cost function (D.40) is equal to $J_0(x_0)$, which is calculated by the last step of the following iteration that proceeds backward in time from period $N-1$ to period 0,

$$J_k(x_k) = \min_{u_k \in U} E_{w_k} \left\{ g_k(x_k, u_k, w_k) + J_{k+1}(f_k(x_k, u_k, w_k)) \right\}, k = N-1, \cdots, 1, 0$$
$$\text{(D.41)}$$
$$J_N(x_N) = g_N(x_N) \qquad \text{(D.42)}$$

and if $u_k^* = \mu_k^*(x_k)$ minimizes the right side of (D.41) for each x_k and k, the policy $\pi^* = \{\mu_0^*, \mu_1^*, \cdots, \mu_{N-1}^*\}$ is optimal.

Stochastic Optimal Control with Infinite Horizon. For the system (D.39) the optimal control problem is for a given initial state x_0 to find a policy $\pi = \{\mu_0, \mu_1, \cdots\}$, $\mu_k : S \to C, \mu_k(x_k) \in U(x_k)$, which minimizes the cost function

$$J_\pi(x_0) = \lim_{N \to \infty} \mathop{E}_{\substack{w_k \\ k=0,1,\cdots}} \left\{ \sum_{k=0}^{N-1} \alpha^k g(x_k, \mu_k(x_k), w_k) \right\} \qquad \text{(D.43)}$$

where α is a positive scalar with $0 < \alpha \leq 1$, called the discount factor. The meaning of $\alpha < 1$ is that future costs matter to us less than the same costs incurred at the present time.

First, it should be noted that this problem involves two assumptions: infinite number of stages and a stationary system; that is, the system equation, the cost per stage, and the random disturbance statistics do not change from one stage to the next. The assumption of an infinite number of stages is never satisfied in practice, but constitutes a reasonable approximation for problems involving a finite but very large number of stages. The assumption of stationary system is often satisfied in practice, and in other cases, it approximates well a situation where the system parameters vary slowly with time.

There are several analytical and computational issues regarding infinite horizon problems. Many of these revolve around the relation between the optimal cost function J^* of the infinite horizon problem and the optimal cost function of the corresponding N-stage problems. Since the infinite horizon cost of a given policy is the limit of the corresponding N-stage cost as $N \to \infty$ by definition, it is natural to speculate the following two issues:

1. The optimal infinite horizon cost is the limit of the corresponding N-stage optimal costs as $N \to \infty$; that is, $J^*(x) = \lim_{N \to \infty} J_N(x)$ for all states x. The optimal N-stage cost is generated after N iterations of the DP algorithm:

$$J_{k+1}(x) = \min_{u \in U(x)} \mathop{E}_{w} \{ g(x, u, w) + \alpha J_k(f(x, u, w)) \}, \quad k = 0, 1, \cdots \qquad \text{(D.44)}$$

    starting from the initial condition $J_0(x) = 0$ for all x.

2. The following limiting form of the DP algorithm should hold for all states x:

$$J^*(x) = \min_{u \in U(x)} E_w\{g(x, u, w) + \alpha J^*(f(x, u, w))\} \qquad \text{(D.45)}$$

which can be viewed as a functional equation for the cost-to-go function J^* and is called Bellman's equation. Furthermore, if $\mu(x)$ attains the minimum in the right-hand side of Bellman's equation for each x, then the policy $\{\mu, \mu, \cdots\}$ should be optimal.

Most of the analysis of infinite horizon problems revolves around the above two issues and also around the issue of efficient computation of J^* and an optimal policy, such as stochastic shortest path problem and discounted problem with bounded cost per stage suitable for both issues 1 and 2, and discounted and undiscounted problems with unbounded cost per stage not suitable for issue 1.

In addition, a principal method for calculating the optimal cost function J^* is called value iteration, that is, the DP iteration

$$J_{k+1}(i) = \min_{u \in U(i)} \left[g(i, u) + \alpha \sum_{j=1}^{s} p_{ij}(u) J_k(j) \right], i = 1, \cdots, s \qquad \text{(D.46)}$$

where $g(i, u)$ is the expected cost of the cost per stage, defined as $g(i, u) = \sum_j p_{ij}(u) \tilde{g}(i, u, j)$, and $\tilde{g}(i, u, j)$ is the cost of using u at state i and moving to state j.

Generally, value iteration requires an infinite number of iterations, although there are important special cases where it terminates finitely. There is an alternative to value iteration, which always terminates finitely. This algorithm is called policy iteration and operates as follows: Start with a stationary policy μ^0 and generate a sequence of new policies $\mu^1, \mu^2, \cdots$, which involves two steps. In the policy evaluation step, given the policy μ^k, compute $J_{\mu^k}(i), i = 1, \cdots, s$, as the solution of the system equations

$$J(i) = g(i, \mu^k(i)) + \alpha \sum_{j=1}^{s} p_{ij}(\mu^k(i)) J(j), \quad i = 1, \cdots, s \qquad \text{(D.47)}$$

in the s unknown $J(1), \cdots, J(n)$. For the policy improvement step, compute a new policy μ^{k+1} as

$$\mu^{k+1}(i) = arg \min_{u \in U(i)} \left[g(i, u) + \alpha \sum_{j=1}^{s} p_{ij}(u) J_{\mu^k}(j) \right], \quad i = 1, \cdots, s \qquad \text{(D.48)}$$

The process is repeated with μ^{k+1} used in place of μ^k, unless $J_{\mu^{k+1}}(i) = J_{\mu^k}(i)$ for all i, in which case the algorithm terminates with the policy μ^k.

Bibliography

[1] S. Kobayashi, S. Plotkin, and S. K. Rinerio. Energy efficiency technology for road vehicles. *Energy Efficiency*, 2(2): 125–137, 2009.

[2] IEEJ. Asian/world energy outlook. Institute of Energy Economics, Japan, November, 2011.

[3] L. Guzzella and C.H. Onder. *Introduction to Modeling and Control of Internal Combustion Engine Systems*. Springer-Verlag, New York, 2004.

[4] L. Eriksson and N. Nielsen. *Modeling and Control of Engines and Drivelines*. Wiley, New York, 2014.

[5] E. Hendricks and S.C. Sorenson. Mean value modeling of spark ignition engines. Detroit, Michigan, SAE, Technical Paper 900616, 1990.

[6] H. Adibi Asl, M. Saeedi, R. Frase, P. Goossens, and J. McPhee. Mean value engine model including spark timing for powertrain control application. Detroit, Michigan, SAE, Technical Paper 010247, 2013.

[7] M. Moran and H.N. Shaprio. *Fundamentals of Engineering Thermodynamics*. Wiley, New York, 1992.

[8] A. Jante. Der Weg zum Wiebe-Brenngesetz. *Krafifahrzeugtechnik*, 9: 340–346, 1961.

[9] M. Yoshida. Thermodynamical analysis of optimum ignition timing: 3rd report, effects of combustion period and heat release diagram. *Transactions of the Japan Society of Mechanical Engineers B*, 64(617): 305–311, 1998.

[10] D. Cho and J.K. Hedrick. Automotive powertrain modeling for control. *Transactions of the ASME: Journal of Dynamic Systems, Measurement and Control*, 111(4): 568–576, 1989.

[11] A. Chevalier, M. Müller, and E. Hendricks. On the validity of mean value engine models during transient operation. Detroit, Michigan, SAE, Techniqual Paper 01-1261, 2000.

[12] A. Balluchi, et al. Automotive engine control and hybrid systems: Challenges and opportunities. *Proceedings of IEEE*, 88(7): 888–912, 2000.

[13] C.F. Aquino. Transient A/F control characteristics of the 5 liter central fuel injection engine. Detroit, Michigan, SAE, Technical Paper 810494, 1981.

[14] G. De Nicolao, R. Scattolini, and C. Siviero. Modelling the volumetric efficiency of IC engines: Parametric, non-parametric and neural techniques. *Control Engineering Practice*, 4(10): 1405–1415, 1996.

[15] M. Thornhill, S. Thompson, and H. Sindano. A comparison of idle speed control schemes. *Control Engineer Practice*, 8(5): 519–530, 2000.

[16] A. Ohata, J. Kako, T. Shen, and K. Ito. Introduction to the benchmark challenge on SICE engine start control problem. In *Proceedings of the 17th World Congress: The International Federation of Automatic Control (IFAC WC)*, Seoul, July 6–11, 2008, pp. 1048–1053.

[17] K.R. Butts, N. Sivashankar, and J. Sun. Application of l_1 optimal control to the engine speed control problem. *IEEE Transactions on Control Systems Technology*, 7(2): 258–270, 1999.

[18] S. Choi and J.K. Hedric. Robust throttle control of automotive engines: Theory and experiment. *Transactions on ASME: Journal of Dynamic Systems, Measurement and Control*, 118(1): 92–98, 1996.

[19] Y. Yildiz, A.M. Annaswamy, D. Yanakiev, and I. Kolmanovsky. Spark-ignition-engine idle speed control: An adaptive control approach. *IEEE Transactions on Control Systems Technology*, 19(5): 990–1002, 2011.

[20] S.D. Cairano, D. Yanakiev, A. Bemporad, I. Kolmanovsky, and D. Hrovat. Model predictive idle speed control: Design, analysis, and experimental evaluation. *IEEE Transactions on Control Systems Technology*, 20(1): 84–97, 2012.

[21] D. Hrovat, D. Colvin, and B.K. Powell. Comments on "Applications of some new tools to robust stability analysis of spark ignition engine: A case study." *IEEE Transactions on Control Systems Technology*, 6(3): 435–436, 1998.

[22] R. Pfiffiner and L. Guzzella. Feedback linearization of a multi-input SI-engine system for idle speed control. In *Proceedings of the 5th IEEE Mediterranean Conference on Control and Systems*, Paphos, Cyprus, 1997, Paper MP4-2.

[23] L.G. Glielmo, S. Santini, and G. Serra. Optimal idle speed control with induction-to-power finite delay for SI engines. In *Proceedings of the 7th Mediterranean Conference on Control and Automation*, Haifa, Israel, 1999, pp. 200–209.

[24] L.G. Glielmo, S. Santini, and I. Cascella. Idle speed control through output feedback stabilization for finite time delay systems. In *Proceedings of the 2000 American Control Conference (ACC)*, Chicago, IL, 2000, pp. 45–49.

[25] X.Q. Li and S. Yurkovich. Sliding mode control of delayed systems with application to engine idle speed control. *IEEE Transactions on Control Systems Technology*, 9(6): 802–810, 2001.

[26] S.C. Bengea, X.Q. Li, and R.A. DeCarlo. Combined controller-observer design for uncertain time delay systems with application to engine idle speed control. *Transactions of the ASME: Journal of Dynamic Systems, Measurement and Control*, 126(4): 772–780, 2004.

[27] J.A. Cook, et al. Automotive powertrain control: A survey. *Asian Journal of Control*, 8(3): 237–260, 2006.

[28] J. Hale and S. Lunel. *Introduction to Functional Differential Equations.* Springer-Verlag, New York, 1993.

[29] A. Stotsky, B. Egardt, and S. Eriksson. Variable structure control of engine idle speed with estimation of unmeasurable disturbances. *Transactions of the ASME: Journal of Dynamic Systems, Measurement and Control*, 122(4): 599–603, 2000.

[30] J.J. Moskwa. Automotive Engine Modeling for Real Time Control. PhD thesis, Massachusetts Institute of Technology, Boston, 1988.

[31] H.K. Khalil. *Nonlinear Systems.* 3rd ed., Prentice Hall, Upper Saddle River, NJ, 2002.

[32] J. Zhang, T. Shen, and R. Marino. Nonlinear speed control scheme and its stability analysis for SI engines. *SICE Journal of Control, Measurement, and System Integration*, 3(1): 43–49, 2010.

[33] J. Zhang, T. Shen, J. Kako, and S. Yoshida. Lyapunov-based feedback design and experimental verification of IC engine speed control. *International Journal of Control, Automation, and Systems*, 7(4): 659–667, 2009.

[34] J. Zhang, T. Shen, and R. Marino. Model-based cold-start speed control scheme for spark ignition engines. *Control Engineering Practice*, 18(11): 1285–1294, 2010.

[35] J. Zhang, T. Shen, J. Kako, and R. Marino. Design and validation of a model-based starting speed control scheme for spark ignition. *Asian Journal of Control.* DOI: 10.1002/asjc.979.

[36] J. Sun, I. Kolmanonsky, J.A. Cook, and J.H. Bucklan. Modeling and control of automotive powertrain systems: A tutorial. In *Proceedings of the 2005 American Control Conference (ACC)*, Portland, OR, 2005, pp. 3271–3283.

[37] J. Sun, Y.W. Kim, and L. Wang. Aftertreatment control and adaptation for automotive lean burn engines with HEGO sensors. *International Journal of Adaptive Control Signal Processing*, 18(2): 145–166, 2004.

[38] P. Andersson and L. Eriksson. Air-to-cylinder observer on a turbocharged SI-engine with wastegate. SAE, Technical Paper 01-0262, 2001.

[39] P. Andersson and L. Eriksson. Cylinder air charge estimator in turbocharged SI-engines. Detroit, Michigan, SAE, Technical Paper 01-1366, 2004.

[40] J. Grizzle, J. Cook, and W. Milam. Improved cylinder air charge estimation for transient air fuel ratio control. In *Proceedings of the 1994 American Control Conference*, Baltimore, MD, June 4–6, 1994, pp. 1568–1573.

[41] H. Kerkeni, J. Lauber, and T. Guerra. Estimation of individual in-cylinder air mass flow via periodic observer in Takagi-Sugeno form. In *Proceedings of IEEE Vehicle Power and Propulsion Conference (VPPC)*, Lille, France, Sep. 1–3, 2010, pp. 1–6.

[42] A. Stotsky and I. Kolmanovsky. Application of input estimation technique to charge estimation and control in automotive engines. *Control Engineering Practice*, 10(12): 1371–1383, 2002.

[43] A. Stotsky, I. Kolmanovsky, and S. Eriksson. Composite adaptive and input observer-based approaches to the cylinder flow estimation in spark ignition automotive engines. *International Journal of Adaptive Control and Signal Processing*, 18(2): 125–144, 2004.

[44] A. Chevalier, C. Vigild, and E. Hendricks. Predicting the port air mass flow of SI engines in air/fuel ratio control applications. SAE 2001 World Congress & Exhibition, Detroit, Michigan, SAE, Technical Paper 01-0260, 2000.

[45] C. Beltrami, et al. AFR control in SI engine with neural prediction of cylinder air mass. In *Proceedings of the American Control Conference*, Denver, CO, June 4–6, 2003, pp. 1404–1409.

[46] Y. Zhao, T. Shen, and X. Jiao. Air-fuel ratio transient control design for gasoline engines based on individual cylinder air charge estimation, Bangkok, Thailand, April 1–3, SAE Technical Paper 2013-01-0102, 2013.

[47] M. Loganathan and A. Ramesh. Experimental studies on low pressure semi-direct fuel injection in a two stroke spark ignition engine. *International Journal of Automotive Technology*, 10(2): 151–160, 2009.

[48] M.A. Franchek, J. Mohrfeld, and A. Osburn. Transient fueling controller identification for spark ignition engines. *Transactions of the ASME: Journal of Dynamic Systems, Measurement and Control*, 128(3): 499–509, 2006.

[49] P. Dickinson and A.T. Shenton. Dynamic calibration of fueling in the PFI SI engine. *Control Engineering Practice*, 17(1): 26–38, 2009.

[50] M. Shahbakhti, et al. A method to determine fuel transport dynamics model parameters in port fuel injected gasoline engines during cold start and warm-up conditions. *Journal of Engineering for Gas Turbines and Power*, 132(7): 074504-1–074504-5, 2010.

[51] Y. Wu, et al. A study of the characteristic of fuel-film dynamics for four-stroke small-scale spark-ignition engines. Detroit, Michigan, SAE, Technical Paper 01-0591, 2009.

[52] S. Miyashita. Dual injection type internal combustion engine. Patent 2006/0144365 A1, 2006.

[53] L.P. Wyszynski, C.R. Stone, and G.T. Kalghatgi. The volumetric efficiency of direct and port injection gasoline engines with different fuels. Detroit, Michigan, SAE, Technical Paper 01-0839, 2002.

[54] S. Mosbach, et al. Dual injection homogeneous charge compression ignition engine simulation using a stochastic reactor model. *International Journal of Engine Research*, 8: 41–50, 2007.

[55] J. Zhang, T. Shen, G. Xu, and J. Kako. Wall-wetting model based method for air-fuel ratio transient control in gasoline engines with dual injection system. *International Journal of Automotive Technology*, 14(6): 867–873, 2013.

[56] I. Arsiea, C. Pianesea, G. Rizzoa, and V. Cioffib. An adaptive estimator of fuel film dynamics in the intake port of a spark ignition engine. *Control Engineering Practice*, 11(3): 303–309, 2003.

[57] J.S. Souder and J.K. Hedrick. Adaptive sliding mode control of air-fuel ratio in internal combustion engines. *International Journal of Robust and Nonlinear Control*, 14(6): 525–541, 2004.

[58] D. Rupp and L. Guzzella. Adaptive internal model control with application to fueling control. *Control Engineering Practice*, 18(8): 873–881, 2010.

[59] Y. Yildiz, A. Annaswamy, D. Yanakiev, and I. Kolmanovsky. Spark ignition engine fuel-to-air ratio control: An adaptive control approach. *Control Engineering Practice*, 18(12): 1369–1378, 2010.

[60] N.E. Kahveci and M.J. Jankovic. Adaptive controller with delay compensation for air-fuel ratio regulation in SI engines. In *Proceedings of the American Control Conference*, Baltimore, Maryland, 2010, pp. 2236–2241.

[61] X. Jiao, J. Zhang, T. Shen, and J. Kako. Adaptive air-fuel ratio control scheme and its experimental validations for port-injected SI engines. *International Journal of Adaptive Control and Signal Processing*, 29(1): 41–63, 2015.

[62] U.M. Ascher, R.M.M. Mattheji, and R.D. Russell. *Numerical Solution of Boundary Value Problems for Ordinary Differential Equations*. SIAM, Philadelphia, PA, 1995.

[63] K.S. Holkar and L.M. Waghmare. An overview of model predictive control. *International Journal of Control and Automation*, 3(4): 47–63, 2010.

[64] J. Kennedy. Particle swarm optimization. In *Encyclopedia of Machine Learning*, Springer, NY, 2010, pp. 760–766.

[65] P.E. Gill, W. Murray, and M.A. Saunders. SNOPT: An SQP algorithm for large-scale constrained optimization. *SIAM Review*, 47(1): 99–131, 2005.

[66] S.L. Lehman and L.W. Stark. Three algorithms for interpreting models consisting of ordinary differential equations: Sensitivity coefficients, sensitivity functions, global optimization. *Mathematical Biosciences*, 62(1): 107–122, 1982.

[67] T. Ohtsuka. A continuation/GMRES method for fast computation of nonlinear receding horizon control. *Automatica*, 40(4): 563–574, 2004.

[68] C.T. Kelley. Iterative methods for linear and nonlinear equations. In *Frontiers in Applied Mathematics*. Vol. 16. SIAM, Philadelphia, PA, 1995.

[69] Y. Saad. *Iterative Methods for Sparse Linear Systems*. 2nd ed. Society for Industrial and Applied Mathematics, NY, 2003.

[70] E.B. Arthur and H. Yu-Chi. *Applied Optimal Control: Optimization, Estimation, and Control*. Blaisdell Publishing, Waltham, MA, 1969.

[71] J. Zhang and X. Jiao. Adaptive coordinated control of engine speed and battery charging voltage. *Journal Control Theory Applications*, 6(1): 69–73, 2008.

[72] Q. Feng, C. Yin, and J. Zhang. A transient dynamic model for HEV engine and its implementation for fuzzy-PID governor. *IEEE International Conference on Vehicular Electronics and Safety*, Xi'an, China, October 14–16, 2005, pp. 73–78.

[73] J.R. Wagner, D.M. Dawson, and Z. Liu. Nonlinear air-to-fuel ratio and engine speed control for hybrid vehicles. *IEEE Transactions on Vehicular Technology*, 52(1): 184–195, 2003.

[74] A. Isidory. *Nonlinear Control Systems*. Springer, London, 1995.

[75] F. Tahir, T. Ohtsuka, and T. Shen. Tuning of nonlinear model predictive controller for the speed control of spark ignition engines. In *Automatic Control Conference (CACS)*, Nantou, Taiwan, December 2–4, 2013.

[76] F. Tahir and T. Ohtsuka. Tuning of performance index in nonlinear model predictive control by the inverse linear quadratic regulator design method. *SICE Journal of Control, Measurement, and System Integration*, 6(6): 387–395, 2013.

[77] T. Fujii and M. Narazaki. A complete optimality condition in the inverse problem of optimal control. *SIAM Journal on Control and Optimization*, 22(2): 327–341, 1984.

[78] A. Jameson and E. Kreindler. Inverse problem of linear optimal control. *SIAM Journal on Control*, 11(1): 1–19, 1973.

[79] T. Fujii. A new approach to the LQ design from the viewpoint of the inverse regulator problem. *IEEE Transactions on Automatic Control*, 32(11): 995–1004, 1987.

[80] Z. Yang, H. Iemura, S. Kanae, and K. Wada. Identification of continuous-time systems with multiple unknown time delays by global nonlinear least-squares and instrumental variable methods. *Automatica*, 43(7): 1257–1264, 2007.

[81] M. Hong, T. Shen, M. Ouyang, and J. Kako. Nonlinear torque observers design for spark ignition engines with different intake air measurement sensors. *IEEE Transactions on Control Systems Technology*, 19(1): 229–237, 2011.

[82] M. Hong, M. Ouyang, and T. Shen. Torque-based optimal vehicle speed control. *International Journal of Automotive Technology*, 12(1): 45–49, 2011.

[83] M. Kang and T. Shen. Nonlinear model predictive torque control for IC Engines. In *Proceedings of the 11th World Congress on Intelligent Control and Automation*, Shenyang, China, June 29–July 4, 2014, pp. 804–809.

[84] D.W. Clarke, C. Mohtadi, and P.S. Tuffs. Generalized predictive control. Part I. The basic algorithm. *Automatica*, 23(2): 137–148, 1987.

[85] E.F. Camacho and C. Bordons. *Model Predictive Control*. Springer-Verlag, London, 1998.

[86] L. Weiss. Controllability, realization and stability of discrete-time systems. *SIAM Journal on Control*, 10(2): 231–251, 1972.

[87] K. Suzuki, T. Shen, J. Kako, and S. Yoshida. Individual A/F estimation and control with fuel-gas ratio for multi-cylinder IC engines. *IEEE Transactions on Vehicular Technology*, 58(9): 4757–4768, 2009.

[88] Y. Liu and T. Shen. Modeling and experimental validation of air-fuel ratio under individual cylinder fuel injection in gasoline engines. *IEEE Transactions on Industry Applications*, 1(3): 155–163, 2012.

[89] Y. Mutoh. A new design procedure of the pole-placement and the state observer for linear time-varying discrete system. In *Informatics in Control, Automation and Robotics*, ed. J.A. Cetto et al. LNEE 89. Springer-Verlag, Berlin, 2011, pp. 321–333.

[90] T. Hara, T. Shen, and Y. Mutoh. Iterative learning-based air-fuel control of gasoline engines with unknown off-set. In *IEEE International Conference on 'Intelligent Systems,'* Sofia, Bulgaria, September 6–8, 2012, vol. I, pp. 304–309.

[91] F. Galliot, W.K. Cheng, C.O. Cheng, M. Sztenderowicz, J.B. Heywood, and N. Collings. In-cylinder measurements of residual gas concentration in a spark ignition engine. Detroit, Michigan, SAE, Technical Paper 900485, 1990.

[92] J.W. Fox, W.K. Cheng, and J.B. Heywood. A model for predicting residual gas fraction in spark-ignition engines. Detroit, Michigan, SAE, Technical Paper 931025, 1993.

[93] G. Tsutomu. Thermal effect upon air capacity of the four-stroke engine (1st report, thermodynamical analysis method), *Bulletin of JSME*, 12(53): 1163–1179, 1969.

[94] P. Giansetti, G. Colin, P. Higelin, and Y. Chamaillard. Residual gas fraction measurement and computation. *International Journal of Engine Research*, 8(4): 347–364, 2007.

[95] T. Leroy, G. Alix, J. Chauvin, A. Duparchy, and F. Le Ber. Modeling fresh air charge and residual gas fraction on a dual independent variable valve timing SI engine. Detroit, Michigan, SAE, Technical Paper 01-0983, 2008.

[96] H. Cho, K. Lee, J. Lee, J. Yoo, and K. Min. Measurements and modeling of residual gas fraction in SI engines. Detroit, Michigan, SAE, Technical Paper 01-1910, 2001.

[97] M. Mladek and C. Onder. A model for the estimation of inducted air mass and the residual gas fraction using cylinder pressure measurements. Detroit, Michigan, SAE, Technical Paper 01-0958, 2000.

[98] C.S. Daw, M.B. Kennel, C.E.A. Finney, and F.T. Connolly. Observing and modeling nonlinear dynamics in an internal combustion engine. *Physical Review E*, 57(3): 2811–2819, 1998.

[99] J. Davis, C.S. Daw, L.A. Feldkamp, J.W. Hoard, F. Yuan, and T. Conolly. Method of controlling cyclic variation engine combustion. U.S. Patent 5,921,221, 1999.

[100] J. Yang, T. Shen, and X. Jiao. Model-based stochastic optimal air-fuel ratio control with residual gas fraction of spark ignition engines. *IEEE Transactions on Control Systems Technology*, 22(3): 896–910, 2014.

[101] J. Yang, T. Shen, and X. Jiao. Stochastic robust air-fuel ratio control with residual gas fraction of spark ignition engines. *Journal of Control Theory and Application*, 11(4): 586–591, 2013.

[102] O.L.V. Costa. Stability result for discrete-time linear systems with Markovian jumping parameters. *Journal of Mathematical Analysis and Applications*, 179(1): 154–178, 1993.

[103] W.K. Ching and K.N. Michael. *Markov Chains Models, Algorithms and Applications*. Springer-Verlag, Berlin, 2006.

[104] J.B. Heywood. *Internal Combustion Engine Fundamentals*. McGraw-Hill, New York, 1988.

[105] E. Hendricks and S.C. Sorenson. Mean value SI engine model for control studies. In *Proceedings of the 1990 American Control Conference (ACC)*, San Diego, CA, 1990, No. TP10-6.

[106] M. Abate and V. Di Nunzio. Idle speed control using optimal regulation. Detroit, Michigan, SAE, Technical Paper 905008, 1990.

[107] T. Jimbo and Y. Hayakawa. Physical-model-based control of engine cold start via role state variables. In *Proceedings of the 17th World Congress: The International Federation of Automatic Control (IFAC WC)*, Seoul, July 6–11, 2008, pp. 1024–1029.

[108] M. Ogawa and H. Ogai. The cold start control of engine using large scale database-based online modeling. In *Proceedings of the 17th World Congress: The International Federation of Automatic Control (IFAC WC)*, Seoul, July 6–11, 2008, pp. 1030–1035.

[109] S. Sgihira, S. Kitazono, and H. Ohmori. Starting speed control of SI engine based on extremum seeking control. In *Proceedings of the 17th World Congress: The International Federation of Automatic Control (IFAC WC)*, Seoul, July 6–11, 2008, pp. 1036–1041.

[110] J. Zhang and T. Shen. Model-based cold-start speed control design for SI engine. In *Proceedings of the 17th World Congress: The International Federation of Automatic Control (IFAC WC)*, Seoul, July 6–11, 2008, pp. 1042–1047.

[111] K. Ropke. DOE-design of experiments method and applications in engine development. SV Corporate Media, 2005.

[112] A. Ohata. A desired modeling environment for automotive controls. In *Identification for Automotive Systems*. Lecture Notes in Control and Information Sciences, vol. 418. Springer-Verlag, London, 2012, pp. 13–34.

[113] J.C. Livengood and P.C. Wu. Correlation of autoignition phenomena in internal combustion engines and rapid compression machines. *Symposium (International) on Combustion*, 5(1): 347–356, 1955.

[114] M. Krstic, I. Kanellakopoulos, and P. Kokotovic. *Nonlinear and Adaptive Control Design*. Wiley, New York, 1995.

[115] S. Sastry. *Nonlinear Systems Analysis, Stability and Control*. Springer, New York, 1999.

[116] V.B. Kolmanovskii and A. Myshki. *Introduction to the Theory and Applications of Functional Differential Equations*. Kluwer Academic, Dordrecht, Netherlands, 1999.

[117] U. Jonsson, C. Trygger, and P. Ogren. *Optimal Control*. Lecture note. Royal Institute of Technology, Stockholm, Sweden, 2010.

[118] J.L. Troutman. *Variational Calculus and Optimal Control: Optimization with Elementary Convexity*, 2nd ed. Springer-Verlag, New York, 1995.

[119] X. Xie. *Optimal Control: Theory and Application*. Tsinghua University Press, Beijing, 1986 (in Chinese).

[120] K. Najim, E. Ikonen, and A.K. Daou. *Estimation, Stochastic Processes, Estimation, Optimization and Analysis*. Butterworth-Heinemann, Sterling, VA, 2004.

[121] S.P. Meyn and R.L. Tweedie. *Markov Chains and Stochastic Stability*. Springer-Verlag, Berlin, 2005.

[122] V. Dragan, T. Morozan, and A.M. Stoica. *Mathematical Methods in Robust Control of Discrete-Time Linear Stochastic Systems*. Springer, Business Media, New York, 2010.

[123] R. Khasminskii. *Stochastic Stability of Differential Equations*. Springer, New York, 2010.

[124] D. Bertsekas. *Dynamic Programming and Optimal Control*. Vol. I. Athena Scientific, Belmont, MA, 1995, ch. 1.

[125] D. Bertsekas. *Dynamic Programming and Optimal Control*. Vol. II. Athena Scientific, Belmont, MA, 1995, ch. 1.

Index